Hemodialysis Membranes
For Engineers to Medical Practitioners

Hemodialysis Membranes
For Engineers to Medical Practitioners

Sirshendu De and Anirban Roy

CRC Press
Taylor & Francis Group
Boca Raton London New York

CRC Press is an imprint of the
Taylor & Francis Group, an **informa** business

CRC Press
Taylor & Francis Group
6000 Broken Sound Parkway NW, Suite 300
Boca Raton, FL 33487-2742

First issued in paperback 2020

ISBN 13: 978-0-367-57373-7 (pbk)
ISBN 13: 978-1-138-03293-4 (hbk)

Library of Congress Cataloging-in-Publication Data

Names: De, Sirshendu, author. | Roy, Anirban, author.
Title: Hemodialysis membranes / Sirshendu De, Anirban Roy.
Description: Boca Raton: Taylor & Francis, 2017. | Includes bibliographical references and index.
Identifiers: LCCN 2016040635| ISBN 9781138032934 (hardback : alk. paper)
Subjects: LCSH: Hemodialysis. | Membranes (Technology) | Blood--Filtration.
Classification: LCC RC901.7.H45 D42 2017 | DDC 617.4/61--dc23
LC record available at https://lccn.loc.gov/2016040635

Visit the Taylor & Francis Web site at
http://www.taylorandfrancis.com

and the CRC Press Web site at
http://www.crcpress.com

Contents

Contents

Contents

Preface

The worldwide epidemic of kidney disease is evident from the increasing number of renal failure patients. This number is increasing at an alarming rate. The effect of renal failure is so pronounced that 10% of the world population is affected by kidney-related diseases and millions lose their precious lives since they cannot afford the right kind of treatment. Hemodialysis, the treatment for chronic kidney failure patients, is an expensive procedure in developing economies like South-East Asia, particularly India. The biggest challenge of hemodialysis technology lies in the science behind the membranes. These membranes have special characteristics for clinical specifications as well as physical requirements. There is a lot of literature presently available that documents this, but very little when it comes to addressing the real science and engineering as well as clarifying the basics of hemodialysis membranes, its requirements, and its limitations.

The present book is an effort in this regard, where the authors have developed the content keeping in mind both nephrologists and medical practitioners along with chemical, materials, and polymer engineers. This book discusses the brief history and clinical parameters of dialysis technology, with state-of-the-art development in both material and technological advancements. With this, the book completes a journey from synthesizing a suitable dialysis membrane material to the spinning of hemodialysis fibers and developing posttreatment strategies. The unique feature of this book is a discussion on not only material science, but also the development of a spinning technology that is of low cost and affordable. Moreover, a pragmatic approach has been adopted keeping in mind business and financial analysis of the developed product in the context of a developing country, viz., India.

The authors believe that this book will help professionals from a wide range of backgrounds in understanding the nuances of hemodialysis as a therapy as well as a technology. It is important to appreciate that in order to mimic nature's evolution spread over millions of years, scientists have delved into a century of research and recent advancements are still in progress. It is indeed challenging to capture all the properties and functionalities of a human kidney in a single dialyzer cartridge, but the human race has slowly but steadily been reaching the ultimate design, in terms of material as well as physical properties.

Preface

The book has maintained a continuity in text and is ideal as an introductory material for membrane as well as biomedical engineering students. The authors would like to thank their family and well-wishers for their unending support and encouragement during the preparation of the book.

Sirshendu De

Anirban Roy

Authors

Dr. Sirshendu De is a professor in the Department of Chemical Engineering, Indian Institute of Technology, Kharagpur (IIT KGP). He completed his undergraduate (1986–1990) and master's degrees (1991–1992) and his PhD (1993–1997), all from IIT Kanpur. Having completed his PhD, he joined IIT KGP as visiting faculty in 1997 and rose to associate professor in 2002, finally becoming a professor in 2007. Professor De is an eminent chemical engineer in India. He maintains national and international visibility through his innumerable international publications (240), patents (15), books (7) and book chapters (10), and three technology transfers. Apart from developing low-cost spinning for both dialysis membranes and hollow fibers for water treatment and other industrial use, some of his other innovations include ultra-low-cost filters for the removal of arsenic from drinking water, extraction of polyphenols from green tea leaves, and processing of packaged tender coconut water with high shelf life without any additive and preservative. Professor De has been awarded with the prestigious Shanti Swarup Bhatnagar (2011), the highest scientific honor in India. He is the recipient of the Innovation Award 2015 from the Indian Desalination Association (south zone); the National Award (Govt. of India) for Technology Innovation in various fields such as the petrochemicals and downstream plastics processing industry (2015); the Department of Science and Technology (Govt. of India), Lockheed Martin Gold Medal, Top 10 Innovation Award (2015); the IC2 Business Development Support, NASI Reliance Award (2013); and the Herdilia Awards (2010) to name a few. He is a fellow of INAE and NASI. Presently, he is INAE chair professor and is serving as the head of the Department of Chemical Engineering.

Dr. Anirban Roy completed his undergraduate training in chemical engineering from the Heritage Institute of Technology, Kolkata, in 2007 and master's in chemical engineering from Jadavpur University, Kolkata, in 2009. He had a brief stint as a research assistant at the University of Rhode Island from 2009 to 2011 and worked as a consultant engineer at M.N. Dastur and Co Pvt. Ltd. for TATA Steel, Jindal, and Vizag Steel. Thereafter, he completed his PhD in the Department of Chemical Engineering, IIT Kharagpur, under the guidance of Prof. Sirshendu De (2012–2016). He has worked extensively on membrane separations specializing in hemodialysis membranes. He was involved in developing the indigenous process design of hemodialysis grade hollow fibers as a topic of his PhD dissertation. He has 12 international journals publications and has 10 conference publications (international and national). He has also filed three Indian and one U.S. patents. Anirban is the recipient of the "Young Innovator of India 2015" award in the Make in India conclave held at IIT Bombay; the National Award (Govt. of India) for Technology Innovation in various fields of petrochemicals and downstream plastics processing industry (2015); Department of Science and Technology (Govt. of India), the Lockheed Martin Gold Medal Top 10 Innovation Award (2015); the IC2 Business Development Support, IChemE Global Award (Top 5 shortlist) for Dhirubhai Ambani Award for Outstanding Chemical Engineering Innovation for Resource-Poor People and Simulanis Research Challenge— Top 3 Innovations in India. His research interests include biomedical applications of membrane technologies, polymer blend thermodynamics, rheology, and clarification of food products and water treatment.

Kidney Function and Uremia

I could prove God statistically. Take the human body alone—the chances that all the functions of an individual would just happen is a statistical monstrosity.

—George Gallup

1.1 Kidney Function and Micturition

The human kidney performs a multitude of functions. Primarily, these functions include[1] the following:

1. *Excretion of metabolic wastes*—Metabolism produces a lot of waste products that need to be excreted from the body, in order to avoid accumulation. A few examples of such waste products are urea, creatinine, and other metabolites. Apart from these, kidneys also eliminate toxins like pesticides, drugs, and artificial food additives that are ingested.[1]

2. *Water and electrolyte balances and body fluid osmomolality*—Homeostasis is the property of a system by virtue of which any external disturbance leading to disruption of a system can be adjusted by controlling specific parameters. Kidneys maintain the homeostasis of the human body by regulating water and electrolyte balance. Intake of water and electrolytes can increase or decrease, at times drastically, and it can also be prolonged. In such cases, the kidney maintains the overall balance of electrolyte and water content by adjusting the excretion rates accordingly.[2]

3. *Arterial pressure regulation*—This is maintained by controlling the water and sodium concentration in the blood. At times, short-term arterial pressure regulation is also carried out by manipulating vasoactive factors like rennin.[2]

4. *Acid–base balance*—The kidney maintains acid–base balance in tandem with the lungs and body fluid buffers. It helps in excreting acids (e.g., sulfuric acid, phosphoric acid) produced during protein metabolism.[2]

5. *Hormonal secretion and balance*—The kidneys produce the active form of 1,25-dihydroxyvitamin D_3, also known as calcitrol. Calcitrol is required for normal phosphate deposition on bones and calcium reabsorption in the gastrointestinal tract.[2]

6. *Gluconeogenesis*—The kidneys synthesize glucose from amino acids during prolonged fasting. This is known as gluconeogenesis.[2]

In this regard, it is essential to understand the functioning of the kidney. A basic cross-sectional view of the human kidney is depicted in Figure 1.1a. In general, impure blood from various organs of the human body is received by the kidneys. The toxins are then removed from the body via transport through nephrons in the form of water-soluble waste stream received by the urinary bladder.

The single unit that determines the kidney's filtration is called a nephron (Figure 1.1b). In fact, the size of the kidneys in various species is often determined

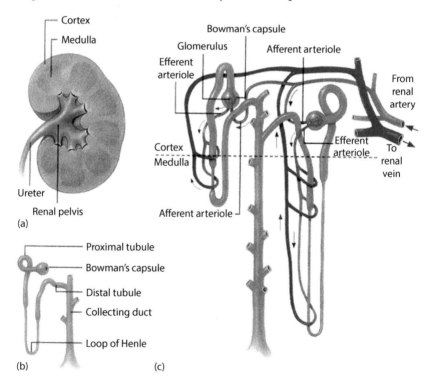

Figure 1.1 (a) Kidney cross section, (b) single nephron, and (c) uremic toxin pathway and urine formation. (Adapted from Kidney: Important Organ of Human Body. http://humanbodyreviews.blogspot.in/, Access date : 2016, April 15.)

by the number of nephrons. The human kidney has 1 million nephrons. There is a gradual decrease in the number of nephrons with age, and after 40 years, the number of nephrons decreases at the rate of 10% every 10 years. Each nephron consists of an individual renal tubule and a glomerulus (Figure 1.1b). The glomerulus is about 150–200 μm in diameter and has a lot of capillaries. They have high hydrostatic pressure (60 mm Hg), and the entire glomerulus is encased in Bowman's capsule. The urine flow path involves flow through the capillaries to Bowman's capsule to the proximal tubule through the loop of Henle to the distal tubule. Urine then enters into the collecting duct. Each collecting duct collects urine from 4000 nephrons that finally enters the urinary bladder (Figure 1.1c). Uremic toxins are expelled by the body, and the process is called urination or micturition. It is basically a two-step process. The first step involves filling up of the bladder until the tension in the walls rises above a threshold level. As a result, the second step is triggered, involving a nervous reflex called micturition reflex, which in layman terms indicates the urge to urinate.

The study of transport of various uremic toxins, as well as other components like ions, acids across various parts of a nephron, is a complicated and interesting phenomenon to understand. At this point, it is important to appreciate that there is reabsorption and secretion across various parts of the nephron maintaining the delicate balance of different components in blood.

1.1.1 Proximal Tubules: Reabsorption and Secretion

There is an appreciable reabsorption of sodium and water in the proximal tubules (65%) before the filtrate reaches the loop of Henle. Proximal tubules have the capacity for active and passive reabsorption due to the high surface area of cells, which is attributed to the presence of an extensive brush border and a huge labyrinth of basal channels. Transport of sodium is carried out by co-transport or counter-transport mechanisms. The exposed surfaces of proximal tubules have protein carrier molecules, helping in major co-transport of sodium along with organic nutrients like amino acids and glucose. The remaining sodium is transported through a counter-transport mechanism involving reabsorption of sodium and secretion of hydrogen ions.[2] Proximal tubules are also responsible for the secretion of organic acids and bases such as bile, salts, oxalates, urate, and catecholamines and para-aminohippuric acid (PAH).[2]

1.1.2 Solute and Water Transport in the Loop of Henle

The loop of Henle has three distinct parts—thin ascending, thin descending, and thick ascending segments. The thin segments have thin epithelial membranes with minimum metabolic activity and conversely the thick segments have thicker epithelial cells with high metabolic activity. Both the ascending limbs (thick and thin) are impermeable to water, and these are responsible for the concentration of urine. The whole of water reabsorption, in the loop of Henle, occurs in the thin descending limb. Sodium, chloride, and potassium are reabsorbed in the thick ascending limb, along with calcium, bicarbonate, and magnesium.

3

1.1.3 Distal Tubule

This part has a very important function in providing the feedback signal to control the filtration rate and regulating blood flow to the nephron. This is through an arrangement known as the juxtaglomerular complex. The rear part of the distal tubule is convoluted and absorbs ions and is impermeable to water.

In order to understand the actual filtration mechanism prevailing in the glomerulus, a closer look into the glomerular capillaries is required. This would reveal that the input bloodstream (impure), to the glomerulus, is through the afferent arteriole and after filtration it is sent back for circulation through the efferent arteriole (Figure 1.2). A closer look at Figure 1.2 would indicate that capillaries have podocytes on the endothelium. The function of podocytes can be fully understood if one delves deeper. This is explored in detail in Figure 1.3.

The glomerular capillaries are fenestrated with 70–90 nm diameter pores on the endothelium. The epithelial cells, or podocytes, have numerous pseudopodia. They interdigitate along the capillary walls to form slits, which are 25 nm wide, each closed by slit diaphragms.[3] These slit diaphragms have surface proteins such as nephrin, podocalyxin, and P-cadherin.[3] These are primarily responsible for restricting the passage of serum albumin and gamma globulins through them, maintaining their composition in the blood. Small

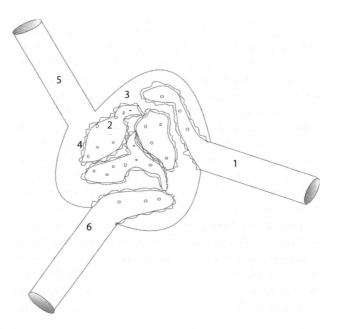

Figure 1.2   Glomerulus of human kidney. (1) Afferent arteriole, (2) red blood cells, (3) Bowman's space, (4) podocyte processes, (5) proximal tubule, and (6) efferent arteriole.

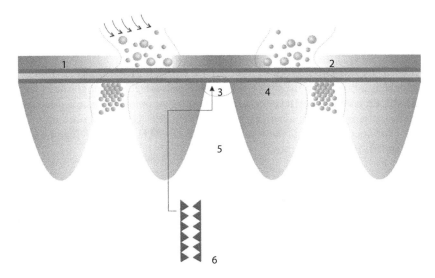

Figure 1.3 Schematic of filtration across the glomerulus. (1) Endothelial cells, (2) pores (70–90 nm), (3) filtration slits (25 nm), (4) podocytes, (5) Bowman's capsule, and (6) diaphragm opening.

molecules pass through the slits along with water. Thus, the basic aim served by this arrangement is to arrest the passage of any neutral solute greater than 8 nm and provide free passage to anything till 4 nm. The passage of solutes is also dependent on the charge of molecules. The total area of filtration, for humans, in the glomerular capillary is about 0.8 m^2. However, this is an average estimate of the filtration area, which can be adjusted by the contraction or expansion of podocytes and, consequently, contraction of the diaphragm slits altering active filtration and affecting blood filtration rate.[2]

1.2 Quantifying Kidney Performance

In order to get an idea about the performance of the human kidney, it is necessary to index it in quantifiable terms.

1.2.1 Effective Renal Plasma Flow

In normal human adults, about 25% of the cardiac output of blood flow received by the kidney is in the order of 1.2–1.3 L/min. Renal plasma flow (RPF) is defined as the volume of plasma that reaches the kidney per unit time. Now, if extraction ratio is defined as

$$\text{Extraction ratio} = \frac{\text{Concentration of compound entering the kidney}}{\text{Concentration of compound excreted into the urine}}$$

$$(1.1)$$

Then, effective renal plasma flow (ERPF) is defined as

$$ERPF = RPF \times \text{Extraction ratio} \tag{1.2}$$

Thus, if the extraction ratio is 1, then RPF is equal to ERPF. Hence, if an indicator that is not metabolized by the body and passes freely through the kidney is used, then measuring its concentration both before it enters the kidney and in the urine would provide an estimate of the ERPF. The most common indicator for such analysis is para-aminohippurate (PAHA). An average human has an ERPF of 625 mL/min.[1]

PROBLEM 1.1

Given concentration of PAHA in urine is 20 mg/L, urine flow is 0.8 mL/min, and concentration of PAHA in plasma is 0.018 mg/L, find (a) ERPF, (b) actual renal plasma flow (RPF), and (c) renal blood flow. Extraction ratio can be assumed as 0.9 and hematocrit (Hct) concentration can be assumed to be 45%.

Solution: (a) Concentration of PAHA in urine (C_{PAHA}) = 20 mg/mL

Urine flow ($\dot{V}$): 0.8 mL/min

Concentration of PAHA in plasma (P_{PAHA}) = 0.018 mg/mL

$$ERPF = \frac{C_{PAHA}\dot{V}}{P_{PAHA}} = 888 \text{ mL/min}$$

(b) $$RPF = \frac{ERPF}{\text{Extraction ratio}} = 987 \text{ mL/min}$$

(c) Renal blood flow $= RPF \times \dfrac{1}{1 - Hct} = 1794.5$ mL/min

1.2.2 Glomerular Filtration Rate

Along with ERPF, another parameter used to quantify the kidney's performance is the glomerular filtration rate (GFR). It is determined by measuring the concentration of a solute that freely passes through the urine and its concentration in the plasma. Thus GFR is expressed as

$$GFR = \frac{\text{Urine concentration of substance} \times \text{Urine flow per unit time}}{\text{Arterial plasma concentration of substance}} \tag{1.3}$$

Inulin, which has a molecular weight of 5200 g/mol, is a nontoxic substance and is not metabolized by the human body. Hence, this meets the criteria for measuring GFR in humans and various other species.[1]

PROBLEM 1.2

Calculate GFR if concentration of inulin in urine and plasma is 40 and 0.3 mg/mL, respectively, and the urine flow rate is 0.8 mL/min.

Solution: Given: Concentration of inulin in urine (C_{inulin}) = 40 mg/mL

Concentration of inulin in plasma (P_{inulin}) = 0.3 mg/mL

Urine flow rate $(\dot{V})$ = 0.8 mL/min

$$GFR = \frac{C_{inulin} \times \dot{V}}{P_{inulin}} = 107 \text{ mL/min}$$

An average normal person has an approximate GFR of 125 mL/min, and this depends on the surface area of filtration. In females, after correction of surface area, GFR is 10% lower than in males. An interesting aspect to note regarding the filtration capacity of the human body is its ability to reabsorb nutrients during the process. A GFR of 125 mL/min would indicate processing of 180 L/day, but urine output is only 1 L/day. This indicates that 99% of the filtrate is reabsorbed by the filtration mechanism.

1.3 Renal Failure: Types and Causes

Two types of renal failures occur in humans: acute kidney injury (AKI) and chronic kidney disease (CKD).[1]

1.3.1 Acute Kidney Injury

It is an abrupt loss of kidney function that develops in a very short time, even within 7 days. The three main reasons for AKI are

1. *Prerenal*: This occurs when blood flow to the kidney is affected. It can be caused due to dehydration, vomiting, and medical complications like diarrhea, heart attack, liver failure, or even blood loss during surgeries. There is no actual damage to the kidneys in prerenal failure and, if the damage is not prolonged, then with proper treatment, the original kidney function can be recovered.

2. *Postrenal*: This occurs when the flow of urine from the kidneys gets affected. This is due to the development of kidney stones, cancer of the urinary tract, or medications taken by patients. Obstruction due to bladder stones, or blood clots, or bladder cancer can cause irregular urination leading to postrenal failure. The treatment procedure would involve removal of blockage, and kidney function is restored in a couple of weeks' time.

3. *Renal*: This occurs when the kidney itself fails to purify the blood at its full capacity. This may happen due to disease of the blood vessel, blood clotting in the kidneys, and glomerulonephritis.

1.3.2 Chronic Kidney Disease

Contrary to AKI, CKD occurs gradually, with the functioning of the kidney decreasing progressively. It can take over several months or years to occur.

All patients having GFR < 60 mL/min are classified under CKD. There are various causes leading to CKD primarily due to diabetes mellitus, hypertension, and glomerularnephritis. There are five stages of CKD based on severity of kidney dysfunction, and each stage is demarcated by specific GFR values:

Stage 1: Mild reduction in GFR (90 mL/min)

Stage 2: Further reduction of GFR to 60–89 mL/min

Stage 3: Moderate reduction of GFR (30–59 mL/min)

Stage 4: Severe GFR reduction to 15–29 mL/min

Stage 5: End-stage renal disease (ESRD) established with GFR < 15 mL/min

CKD presents a high degree of risk for occurrence of cardiovascular diseases and overall risk of cancer is increased as well.[1,2]

1.4 Kidney Failure: Uremic Toxicity

The precursor to toxin buildup in the human body is kidney malfunction, and this clinical syndrome is referred to as uremia resembling systemic poisoning. There is an overall decline in the performance of various organs of the human body. This can manifest itself through decline in the performance of the cardiovascular system (hypertension, hypotension, etc.), the nervous system (loss of memory, sleep disorders), the hematological system (anemia, bleeding), and the immunological system (susceptible to cancer and infection). Various other complications such as bone diseases (amyloidosis, osteoporosis), skin diseases, gastrointestinal diseases (gastritis, nausea), weight loss, and hypothermia can also arise due to uremia.

Under normal functioning conditions, a human kidney clears toxins up to 58 kDa.[4] In this light, it is important to understand the types of uremic toxins causing specific side effects. This information is of extreme help when it comes to diagnosis and prognosis. The first step involved in this process would be identification of uremic toxins and their respective molecular weights. In fact, uremic toxins are classified according to their molecular weights. The low-molecular-weight toxins are up to 300 Da, and the middle-molecular-weight solutes consist of 300 Da to 15 kDa molecules. Uremic toxins can also be subdivided into protein-bound and non-protein-bound solutes. Simple examples are urea (non-protein-bound) and hippuric acid (protein-bound). Urea is a small, water-soluble molecule, whereas hippuric acid behaves like middle molecules due to protein binding.[4]

1.4.1 Methylguanidine (Molecular Weight 73 Da), Guanidine (Molecular Weight 59 Da)

Guanidines comprise a large group of structural metabolites of arginine. Creatinine also falls under this group but is discussed later. Recently discovered moieties such as asymmetrical and symmetrical dimethylarginine (ADMA and SDMA) also fall in this category. Injection of methylguanidine in dogs resulted in heavy intoxication followed by vomiting and nausea. In rats, in vitro studies suggested that methylguanidines and other guanidines inhibited sodium–potassium ATP-ase.[5] However, overall it was seen that this particular group of toxins was harmful at concentrations much higher than those obtained in uremic patients. Dialytic removal of ADMA is only 20%–30%, in spite of low molecular weight, and this can be explained on the basis of protein binding.[6]

1.4.2 Urea (Molecular Weight 60 Da)
Urea is well reported to cause vomiting, headache, and fatigue in renal failure patients. In guinea pigs, it has been reported to decrease cardiac output.[7] Urea is toxic at high concentrations, but these are higher than those obtained in treated renal failure patients. Hence, urea alone is not to be blamed for the uremic symptoms observed in renal failure patients.

1.4.3 Indoxyl Sulfate (Molecular Weight 251 Da)
This belongs to the group known as indoles. Indoxyl sulfate, tryptophan, melatonin, and indole-3-acetic acid are all indoles. Indoxyl sulfate enhances drug toxicity, inhibiting active tubular secretion and deiodination.[8,9] It is also suspected to be one of the agents behind glomerular sclerosis,[10] affecting the function of cellular organic acid. Due to protein binding, the percentage clearance of indoxyl sulfate is greatly reduced and can be as low as 0%–20%.

1.4.4 Myoinsitol (Molecular Weight 180 Da)
This is reported to cause peripheral neuropathy. It also inhibits the formation of Schwann cells.[11]

1.4.5 Hippuric Acid (Molecular Weight 179 Da)
Hippuric acid interferes with various biochemical functions, for example, the functioning of renal tubules, the choroid plexus of the brain, the ciliary body of the eye to name a few. Hippuric acid reduces drug–protein binding,[12,13] and studies indicate that it interferes with glucose tolerance and platelet cyclo-oxygenase activity.[14] It also inhibits renal tubular transport of organic substances.[15]

1.4.6 Peptides
Peptides are typical middle-molecular-weight toxins. They are known to disrupt biochemical functions and inhibit lymphocyte stimulation. Granulocyte inhibiting protein (molecular weight 28 kDa), degranulation inhibiting protein (molecular weight 24 kDa), atrial natriuretic peptide (molecular weight 3.1 kDa), and neuropeptide Y (molecular weight 4.3 kDa) are some of the peptides responsible for uremia.[16–18]

1.4.7 Parathormone (Molecular Weight 9225 Da)

These are reported to inhibit erythropoiesis, red cell osmotic resistance, and impaired contractility of red blood cells. Excess of parathormone gives rise to intracellular calcium leading to malfunction of almost every organ in the human body.[19]

1.4.8 Beta2 Microglobulin (Molecular Weight 12,000 Da)

Carpal tunnel syndrome is characterized with local deposition of beta2 microglobulin (B2M). Its toxic action surfaces only after prolonged exposure. It is the root cause behind the syndrome known as dialysis-related amyloidosis (DRA) that starts after prolonged dialysis treatment (several years) or in aged dialysis patients.[20]

1.4.9 Spermine (Molecular Weight 202 Da)

It is a polycationic polyamine negatively influencing erythropoiesis and other polyamines such as spermidine, putrescine, and cadaverine, which have a high affinity for cells and proteins. They cause anorexia, vomiting, seizures, and hypothermia.[21,22]

1.5 Conclusion: Treatment for Renal Failure Patients

The kidney is responsible for performing a multitude of functions in the human body. However, it has its own limitations. The tubules making up the nephron unit may get damaged during kidney disease. It has its own recovery mechanism, but in case of severe damage, the tubules may never recover to come back to their original state, thereby causing the kidney to malfunction. The kidney cannot make new nephrons and hence its regeneration is limited. Thus, it is imperative that in case of renal failure, a mechanism be devised to reject the uremic toxins from the body precluding its buildup. Hence, hemodialysis was developed as a method of treatment for renal failure patients. It took over a century of continuous discovery and research and development and the technology continued to evolve in order to yield optimum performance. The challenge in kidney failure and its related treatment lies in mimicking the continuous performance of the intelligent, natural organ with a discrete treatment using an artificial substitute. In this regard, it is important to understand the development of dialysis technology, which ensues in the next chapter.

References

1. Hall, J.E. 2010. *Guyton and Hall Textbook of Medical Physiology.* Elsevier Health Sciences, Philadelphia, PA.
2. Ganong, W.F. and Barrett, K.E. 1995. *Review of Medical Physiology.* Appleton & Lange, Norwalk, CT.

3. Wickelgren, I. 1999. Cell biology: First components found for key kidney filter. *Science* 286: 225–226.

4. Bellomo, R., Ronco, C., Kellum, J.A., Mehta, R.L., and Palevsky, P. 2004. Acute renal failure–definition, outcome measures, animal models, fluid therapy and information technology needs: The Second International Consensus Conference of the Acute Dialysis Quality Initiative (ADQI) Group. *Crit. Care* 8(4): R204.

5. Minkoff, L., Gaertner, G., Darab, M., Mercier, C., and Levin, M.L. 1972. Inhibition of brain sodium-potassium ATPase in uremic rats. *J. Lab. Clin. Med.* 80: 71.

6. MacAllister, R.J., Rambausek, M.H., Vallance, P., Williams, D., Hoffmann, K.H., and Ritz, E. 1996. Concentration of dimethyl-L-arginine in the plasma of patients with end-stage renal failure. *Nephrol. Dial. Transplant.* 11: 2449–2452.

7. Johnson, W.J., Hagge, W.W., Wagoner, R.D., Dinapoli, R.P., and Rosevear, J.W. 1972. Effects of urea loading in patients with far-advanced renal failure. *Mayo Clin. Proc.* 47: 21–29.

8. MacNamara, P.J., Lalka, D., and Gibaldi, M. 1981. Endogenous accumulation products and serum protein binding in uremia. *J. Lab. Clin. Med.* 98: 730–740.

9. Lindup, W.E., Bischop, K.A., and Collier, R. 1986. Drug binding defect of uraemic plasma: Contribution of endogenous binding inhibitors. In: *Protein Binding and Drug Transport*, Tillement, J.P. and Lindenlaub, E. (eds.). Schattauer Verlag, Stuttgart, Germany.

10. Motojima, M., Nishijima, F., Ikoma, M., Kawamura, T., Yoshioka, T., Fogo, A.B., Sakai, T., and Ichikawa, I. 1991. Role for 'uremic toxin' in the progressive loss of intact nephrons in chronic renal failure. *Kidney Int.* 40: 461–469.

11. Porter, R.D., Cathcart-Rake, W.F., Wan, S.H., Whittier, F.C., and Grantham, J.J. 1975. Secretory activity and aryl acid content of serum, urine, and cerebrospinal fluid in normal and uremic man. *J. Lab. Clin. Med.* 85: 723–733.

12. Gulyassy, P.F., Bottini, A.T., Stanfel, L.A., Jarrard, E.A., and Depner, T.A. 1986. Isolation and chemical identification of inhibitors of plasma ligand binding. *Kidney Int.* 30: 391–398.

13. MacNamara, P.J., Lalka, D., and Gibaldi, M. 1981. Endogenous accumulation products and serum protein binding in uremia. *J. Lab. Clin. Med.* 98: 730.

14. Dzurik, R., Spustova, V., and Gerykova, M. 1987. Pathogenesis and consequences of the alteration of glucose metabolism in renal insufficiency. In: *Uremic Toxins*, Ringoir, S., Vanholder, R., and Massry, S. (eds.). Plenum Publ Co, New York.

15. Cathcart-Rake, W., Porter, R., Whittier, F., Stein, P., Carey, M., and Grantham, J. 1975. Effect of diet on serum accumulation and renal excretion of aryl acids and secretory activity in normal and uremic man. *Am. J. Clin. Nutr.* 28: 1110–1115.

16. Lipkin, G.W., Dawnay, A.B., Harwood, S.M., Cattell, W.R., and Raine, A.E. 1997. Enhanced natriuretic response to neutral endopeptidase inhibition in patients with moderate chronic renal failure. *Kidney Int.* 52: 792–801.

17. Paniagua, R., Franeo, M., Rodriguez, E., Sanchez, G., Morales, G., and Herrera-Acosta, J. 1992. Impaired atrial natriuretic factor systemic clearance contributes to its higher levels in uremia. *J. Am. Soc. Nephrol.* 2: 1704–1708.

18. Ottosson-Seeberger, A., Lundberg, J.M., Alvestrand, A., and Ahlborg, G. 1997. Exogenous endothelin-1 causes peripheral insulin resistance in healthy humans. *Acta Physiol. Scand.* 161: 211–220.

19. Malachi, T., Bogin, E., Gafter, U., and Levi, J. 1986. Parathyroid hormone effect on the fragility of human young and old red blood cells in uremia. *Nephron* 42: 52–57.

20. Vandenbroucke, J.M., Jadoul, M., Maldague, B., Juaux, J.P., Noel, H., and Van Ypersele de Strihou, C. 1986. Possible role of dialysis membrane characteristics in amyloid osteo-arthropathy. *Lancet* 1: 1210–1211.
21. Radtke, H.W., Rege, A.B., La Marche, M.B., Bartos, D., Bartos, F., Campbell, R.A., and Fischer, J.W. 1981. Identification of spermine as an inhibitor of erythropoiesis in patients with chronic renal failure. *J. Clin. Invest.* 67: 1623.
22. Campbell, R.A. 1987. Polyamines. In: *Uremic Toxins*, Ringoir, S., Vanholder, R., and Massry, S. (eds.). Plenum Publ Co, New York.

2

Evolution of Hemodialysis Technology

The human body experiences a powerful gravitational pull in the direction of hope. That is why the patient's hopes are the physician's secret weapon. They are the hidden ingredients in any prescription.

—**Norman Cousins**

2.1 Hemodialysis Technology

Discussions in Chapter 1 reveal that filtration of uremic toxins from the blood plasma occurs across the fenestrations of the glomerular capillaries. These fenestrations are round or ovular transcellular holes selectively transporting toxins across barriers. Thus, the filtration mechanism basically involves the transport of solutes across a semipermeable barrier. In case of kidney failure, it is essential to replace the kidney with a device that has filtration capabilities like the organ itself. In this regard, membranes, synthetic or natural, play an important part. Understanding membranes and developing them for biomedical applications, such as hemodialysis, are an active and challenging area of interdisciplinary research. Thus, in order to appreciate the technological marvel which is the "artificial kidney", it is very important to discuss about the history of its development.

2.1.1 The First Artificial Kidney and Subsequent Development

Thomas Graham (1805–1869) was the first to demonstrate the effectiveness of a vegetable parchment as a selective barrier for species transport. He stretched a vegetable parchment over a wooden gutta-percha hoop and first filled the hoop with a colloidal solution and then with urine.[1] He floated the apparatus

13

over water and found that solutes passed through the parchment into the water.[1] He coined the term *dialysis* for the observed phenomenon.

However, the first apparatus for blood dialysis did not come into existence until 1913. John J. Abel, Rowntree, and Turner from Johns Hopkins University, USA, came up with their *vividiffusion* apparatus.[2] This consisted of celloidin tubes 8 mm in diameter (Figure 2.1) and 40 cm long. Their most efficient apparatus had 32 celloidin tubes and was suitable for animals weighing more than 20 kg.[3] However, their experiments on dogs led to a couple of modifications that paved the way for future development of the device. One was the flattening of tubes, which they predicted would increase efficiency, which in turn led to future coil dialyzers, and the second was the use of very small tubes resembling present-day hollow-fiber membranes.[4,5]

However, the first human dialysis was performed by George Haas (1886–1971) in 1924.[6,7] His technology was developed independently of Abel–Rowntree–Turner because the First World War did not allow exchange of literature between the Allied and Axis powers. His equipment was very similar to that of the Johns Hopkins group and he prepared 1.2 m long tubes, all put together to yield a surface area of 1.5–2.1 m.[2,7] An important note has to be made at this juncture about the preparation of the anticoagulant. Heparin was not available in those days and hence the success of dialysis pivoted around the availability of an effective blood anticoagulant. The group at Johns Hopkins used hirudin.[4] Hirudin was extracted in a solution from the crushed heads of leeches.[4] In fact, the First World War inhibited John Abel's research since leeches were imported from Europe (Hungary), and the war prohibited the import of "foreign objects." Hence, the leeches ordered by John Abel were left to die in Copenhagen.[4] Later, George Haas discovered an effective method to extract hirudin, and by employing his device, he carried out the first dialysis in autumn 1924. The dialysis lasted only 15 minutes with no reports of complications. Thereafter, a terminal uremia patient was dialyzed (the second

Figure 2.1   Vividiffusion apparatus of Abel, Rowntree, and Turner.

hemodialysis in a human being) in February 18, 1925. It lasted 35 minutes. Four more dialyses were also performed later in 1926, with a duration of 30–60 minutes.[8,9] Finally, on January 13, 1928, for the first time George Haas performed a dialysis on a 55 kg man using a dialyzer with an area of 1.5 m^2 and with highly purified heparin as the anticoagulant.[10]

Following all these developments, in the 1930s, Willem Johan Kolff at Groningen University Hospital started thinking about an apparatus with the potential to replicate renal function (Figure 2.2). He was exposed to the wonders of "cellophane" as a semipermeable barrier by Dr. Brinkman. He carried out experiments using cellophane to transport urea selectively first from saline solution and then from blood. However, in the 1940s, due to the invasion of the Axis armies and the suicide of his professor of medicine (since he was a Jew), Kolff moved to a small town called Kampen. There, with the help of engineer Mr. Hendrik Berk, he built a dialyzer that was basically a rotating drum filtration unit with a large surface area and with cellophane as the separating barrier.[11]

Kolff's team carried out a series of 12 dialyses with very few promising results. They were confronted with various complications ranging from hemolysis and hemorrhage to bloodline disconnections, ultimately leading to failure of the procedure. In 1946, Nils Alwall, in Sweden, developed the first dialyzer (Figure 2.3) that could control the filtration rate.[12,13] He also used cellophane tubes as filtration units. Around the same time, again ignorant about the work by Kolff and Alwall due to the Second World War, Murray, Delorme, and Thomas in Toronto, Canada, developed a static coil artificial kidney dialyzer.[14] In 1947, Bodo von Garrelts, in Stockholm, Sweden, developed a stationary coil-type dialyzer (Figure 2.4).[15] Later in 1953, Inouye and Engelberg, in Philadelphia, built a pressure-cooker-type

Figure 2.2 Willem Kolff's rotating drum dialyzer. (From Agar, J. et al., Kolff-Brigham dialysis machine: 1948, n.d., http://www.homedialysis.org/home-dialysis-basics/machines-and-supplies/dialysis-museum, retrieved November 24, 2016.)

Figure 2.3 Alwall dialyzer. (From Twardowski, Z.J., *Hemodial. Int.*, 12(2), 173, 2008. With permission.)

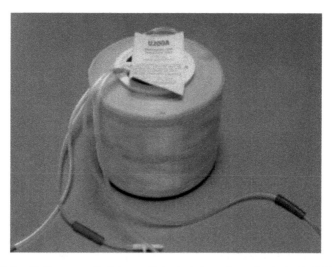

Figure 2.4 Stationary coil dialyzer. (Adapted from Agar, J. et al., Kolff-Brigham dialysis machine: 1948, n.d., http://www.homedialysis.org/home-dialysis-basics/machines-and-supplies/dialysis-museum, retrieved November 24, 2016.)

Figure 2.5   Pressure cooker dialyzer. (Adapted from Agar, J. et al., Kolff-Brigham dialysis machine: 1948, n.d., http://www.homedialysis.org/home-dialysis-basics/machines-and-supplies/dialysis-museum, retrieved November 24, 2016.)

dialyzer (Figure 2.5).[16] The important aspect to note is that while the geometries (and consequently the flow pattern) differed from one dialyzer to another, essentially the separating barrier, or in other words, the membrane, was cellophane or some variant of it.

However, Frederick Kiil described a dialyzer in 1960, which was later modified (Figure 2.6). The modified Kiil dialyzer for the first time replaced cellophane as the membrane and used a less thin, more permeable Cuprophane membrane.[17,18]

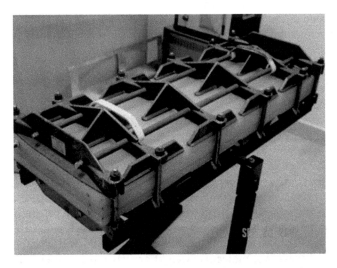

Figure 2.6   Modified Kiil dialyzer. (From Twardowski, Z.J., Hemodial. Int., 12(2), 173, 2008. With permission.)

Scribner shunts deserve special mention in a discussion on the evolution of dialysis technology. They were developed by Dr. Belding Hibbard Scribner. Earlier, Alwall had developed a shunt device made of glass while he was developing his version of the dialyzer. Later, he collaborated with Swedish businessman Holger Crafoord to establish Gambro, a company which has been in the hemodialysis business for the better part of the last five decades. However, Dr. Scribner devised a shunt that later proved a boon for dialysis patients. It was like the shunt device developed by Alwall but with significant modifications and was made of teflon. The shunt was connected to a short piece of elastomeric tubing and was called the Quinton–Scribner shunt (after surgeon Wayne Quinton who collaborated with Dr. Scribner). It was placed on the distal extremity of the leg, between the posterior tibial artery and the anterior saphenous vein. After dialysis sessions, a U-shaped teflon tube was used to shunt the blood flow from the tube in the artery to the tube in the vein. However, from the point of view of membrane engineering, the biggest constraint untill the late 1950s and early 1960s was the synthesis of a proper membrane material for dialysis. Around the same time, another development took place that changed the complete scenario for separation technologies and had a cascading effect in the field of hemodialysis. This was the synthesis of reverse osmosis (RO) membranes by Loeb and Sourirajan utilizing a novel phase inversion technique.[19,20]

2.2 Membrane Technology

It is impossible to comprehend hemodialysis technology completely without understanding membranes. Membranes form an integral and the most important aspect in hemodialysis. The field of membrane separation experienced a quantum leap as soon as Loeb and Sourirajan devised a new and innovative method to synthesize membranes. The method came to be known as the phase inversion technique.

2.2.1 Membrane Synthesis
2.2.1.1 Polymer and Solvent

The first step involved in membrane synthesis is the selection of a polymer. Polymers consist of thousands of structural units called monomers. They are characterized by a chain structure, but their geometries vary from linear to branched, network, star, comb, and so on. Since polymerization is a random process involving chains of nonuniform size, the molecular weight of polymers is not constant but is represented by statistical averaged values. In this regard, polydispersity plays an important role in representing the molecular weight of polymers. Weight, average molecular weight, number average relative molecular weight, Z average molecular weight, and viscosity average molecular weight are various representative molecular weights of polymers.[21] Commercial polymers, more often than not, contain impurities, stabilizers, fillers, plasticizers, and so on. This results in properties of polymers to vary across batches, resulting in differences in the properties of synthesized membranes and the lack of consistency in the results.[22]

For membrane synthesis, the polymer has to be dissolved in a solvent. The solubility of a given polymer in various solvents depends on the Flory–Huggins polymer-solvent interaction parameter (χ_{23}). This is given by the following equation:

$$\chi_{23} = \frac{V_2 \left(\delta_2 - \delta_3 \right)^2}{RT} + \beta \tag{2.1}$$

where

V_2 is the molar volume of solvent

R is the universal gas constant [8.314 J/(mol K)]

T is the temperature in K

δ is the solvent parameter with subscripts 2 and 3 indicating solvent and polymer, respectively

β is an empirical constant

The unit of δ is $(cal\ cm^{-3})^{-0.5}$ and is called the Hildebrand parameter.

2.2.1.2 Lacquer or Casting Solution

The polymer(s)–solvent system is called a lacquer or a casting solution. Lacquers represent concentrated polymeric solutions (5–25 wt%).[22,23] As mentioned by Zeman and Zydney,[22] most of the literature (thermodynamics, rheology, etc.) has been developed for dilute polymeric systems and little literature exists for dense systems. In this regard, it is important to mention the complications arising in dense systems. Polymer solutions can be divided on the basis of concentration as extremely dilute, dilute, semidilute, concentrated, and highly concentrated. In this context, it is important to understand the dynamic contact concentration, c_s. It is defined as the concentration where the polymer chain coils start to "feel" the presence of other coils and begin to interact. At any concentration below c_s, the polymeric chains do not interact and move freely independent of other coils. If c is the concentration of the polymeric solution, then extreme dilute solution is defined as $c < c_s$ (generally between 10^{-2} and 10^{-1} wt%). Normal dilute solution is defined as $c_s < c < c^*$. The variable c^* is defined as the contact concentration, where chains start contacting each other and form multichain aggregations due to translational and rotational motion.[21] It is also referred to as the critical concentration with an order of 10^{-1} wt%. P.G. de Gennes[21] defined c^* as

$$c^* \propto \frac{N}{h^3} = b^{-3} N^{-4/5} \tag{2.2}$$

where

N is the average number of segments corresponding to the average relative molecular weight

$\overline{h}$ is the mean square to end distance

b is the equivalence length of the monomer unit

As the concentration increases beyond c^*, the entanglement length is reached (c^e). Beyond this concentration, chains begin to entangle with each other. In certain cases, if these are sufficiently strong, they can become virtual cross-links, thereby increasing the viscosity of the lacquer. Typically, high-concentration polymer solutions exhibit shear thinning behavior, where their viscosity decreases as shear rates are increased.[22]

Industrially, lacquers are prepared in large reactors. Precise control of shear induced mixing, time of mixing and temperature has to be maintained. This lacquer preparation can extend from hours to several days.[22] After preparation, the lacquer is discharged through micron-sized filters and degassed for removal of gas bubbles. In a typical process, a 10-gallon reactor can yield 36 kg of lacquer.[22] The lacquer is then used to synthesize membranes.

2.2.1.3 Phase Inversion and Membrane Formation

Matsuura[24] has described the complete process of Loeb–Sourirajan membrane synthesis in detail due to its historical importance. The standard Loeb–Sourirajan phase inversion technique is used to prepare asymmetric membranes. In this regard, a brief discussion regarding membrane casting is required. Figure 2.7a depicts the various details of membrane casting showing the steps involved. It involves casting the lacquer over a thin porous fabric forming the support for the membrane and finally immersing this cast film over the fabric in a gelation bath (typically water). In the final step, the membrane is formed, when the water transports across the film and the solvent transports out to the bulk water. This mass transport is depicted in Figure 2.7b. The polymer (P in the figure) stays in the membrane matrix; however, the solvent (S) diffuses into the gelation bath and the non-solvent (NS) in the gelation bath diffuses into the membrane matrix. These transport mechanisms again are determined by the chemical potential gradients. The inter-diffusions lead to the formation of "pores" in the film, thus forming a membrane; intelligent selection of the polymer, solvent, and additives and the temperature and composition of the gelation bath determine the sizes of the pores.

Figure 2.7c depicts the thermodynamic interpretation of this phase inversion mechanism. The phase diagram is a triangle with the pure polymer, pure solvent, and pure NS at the three vertices. Any composition inside the triangle is a mixture of the three. Point A depicts the initial membrane concentration, which on immersing in water undergoes phase inversion. Point A lies outside the binodal curve that joins point P′ to the solvent NS axis. This binodal curve can be obtained from cloud point or simple titration experiments. Dilute polymeric solutions of various concentrations are prepared and distilled water is then added and the solution stirred. The point where the solution turns turbid and turbidity persists is considered as the cloud point. Various compositions of solvent, NS, and polymer can thus be plotted to get the binodal curve. As mentioned earlier, the solvent diffuses out and the NS diffuses in. Point B is the first point where precipitation occurs. Point C is the point when enough solvent has diffused out, yielding a very high viscous phase that can be regarded as solid, which is therefore referred to as point of solidification. Finally, the matrix reaches point D, which yields the final membrane composition. At this

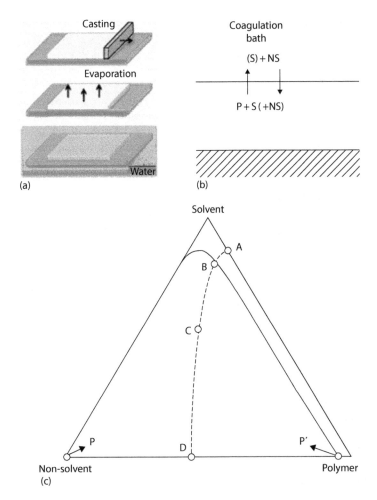

Figure 2.7 Details of membrane formation: (a) steps involved in membrane casting, (b) mass transfer during phase inversion process, and (c) phase diagram showing phase inversion mechanism during membrane formation; A, original casting solution; B, point of precipitation; C, point of solidification; D, final membrane composition, P, polymer lean phase, P′, polymer-rich phase.

point, the two phases are in equilibrium, the polymer lean phase (P), which comprises the pores, and the polymer-rich phase (P′), which forms the membrane structure, with D determining the overall porosity of the membrane. This complete process is often referred to as "demixing."

2.2.2 Membrane Types
Membranes are classified according to their pore sizes as discussed in the following sections.

21

2.2.2.1 Reverse Osmosis

This process involves membranes with the smallest pore size of 2–10 Å, which is sometimes termed as impervious membrane. A good RO membrane separates monovalent salt (sodium chloride) beyond 95%. The transmembrane pressure (TMP) drop required for this process is between 25 and 40 atm and even more in some cases. Permeation is the main transport mechanism in RO. It involves three steps, namely, the dissolution of solutes from the feed stream into the polymeric membrane matrix, the diffusion of solutes through the membrane matrix by diffusion, and desorption from the membrane to the permeate stream. Desalination of sea water is an example of RO.

2.2.2.2 Nanofiltration

Nanofiltration (NF) is closer to RO in its performance but is relatively more porous. It results in partial retention (65%–80%) of monovalent salt (NaCl). The average pore size of the NF membrane is between 5 and 20 Å. The TMP requirement for NF is less, and it is typically between 15 and 25 atm. Solutes of molecular weight between 200 and 1000 Da are separated by NF. The transport mechanism is mainly permeation. Dyes and smaller-molecular-weight organic substances, such as polyphenols, are separated by NF.

2.2.2.3 Ultrafiltration

In this case, the pore size of the membrane is slightly larger, between 20 and 1000 Å. The pressure requirement is significantly lesser as the pore size is much bigger. Six-to-eight atmospheres is the typical TMP drop. Both convection and diffusion are major solute transport mechanisms. The typical molecular weight of solutes to be separated by UF is in the range of 1,000–100,000 Da. Filtration of proteins, polymers, and polysaccharides like pectin can be done using UF.

2.2.2.4 Microfiltration

The average pore size is beyond 1000 Å and up to several microns in microfiltration (MF). The TMP drop is about 2–4 atm. The molecular weight of solutes to be separated by MF is more than 100 kDa. Convection is the main transport mechanism. Separation of blood cells, clay, paint, bacteria, and so on can be achieved by MF. Hemodialysis membranes are essentially UF membranes.

Membranes can be synthesized into two basic geometries: flat sheet and hollow-fiber membranes. Figure 2.8a depicts a typical cross section of a flat sheet membrane. The separation characteristic of membranes depends on the skin layer. This is the thinnest layer of the membrane and, as suggested by Zeman and Zydney,[22] is a morphological discontinuity. It indicates that the mechanism of skin formation is distinctly different from what occurs across the rest of the cross section. The reasons behind it can be (1) the increase in polymer concentration in the top layer induced by desolvation, (2) solid–liquid or liquid–liquid demixing, or (3) polymer adsorption at air–liquid or solid–liquid interface. There are macrovoids visible in the cross section of

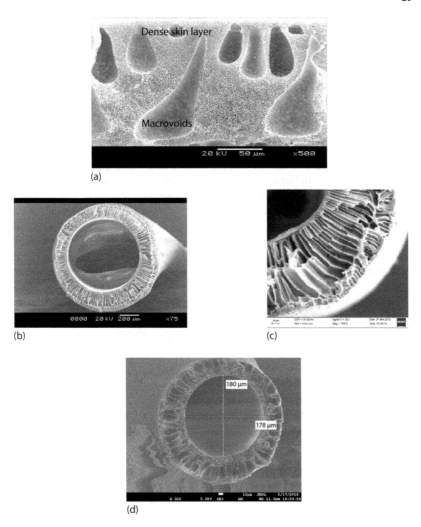

Figure 2.8 (a) Cross section of a flat sheet membrane, (b) cross section of a hollow-fiber membrane, (c) thickness of a hollow-fiber membrane, and (d) cross section of a dialysis fiber membrane.

the membrane, and these are essentially defects in the membrane structure. However, the shape of macrovoids often indicates the nature of demixing during the phase inversion process. The pores are visible as small perforations. Figure 2.8b shows the cross section of the hollow-fiber membrane, with similar morphologically constitutive features. The only variant is the geometry, whereby the feed flows through the inner core and filtration occurs across the radial direction. The magnified view of the wall of the hollow fiber is shown in Figure 2.8c.

2.3 Hemodialysis: State of the Art and Basics
2.3.1 Basic Process

Presently, hemodialysis is carried out using hollow-fiber membranes. These are specialized, clinical grade membranes having an inner diameter of 180–220 μm and a thickness of 35–40 μm. The typical cross section of a hemodialysis membrane is depicted in Figure 2.8d. Seven thousand to ten thousand of such fibers are bundled together (yielding an area of 1.1–1.5 m²) to form a cartridge. This cartridge is placed in an extracorporeal circuit and filters the blood. Figure 2.9 shows the hemodialysis technology employed for renal failure patients. Blood is withdrawn from the body using a pump, which then flows through the fibers in the cartridge. Toxin transport occurs across the membranes due to (1) the concentration gradient between the feed and the dialysate side, (2) the pressure differential applied across the feed and dialysate leading to convective transport, and (3) a combination of (1) and (2). In fact, (1) is classical hemodialysis, whereas (2) is hemofiltration (HF) and (3) is hemodiafiltration (HDF) and is discussed as follows: The toxins are washed away by the dialysate flowing outside the fibers. The circuit is maintained at a continuous flowing condition, and blood coagulation is prevented by periodic injections of heparin. The filtered blood is transferred back to the patient's body. At this juncture, it is important to understand terminologies related to hemodialysis and understand how these parameters help nephrologists quantify the quality of a dialysis operation.

2.3.2 Dialyzer and Dialysate

It is necessary to discuss two fundamental components of the extracorporeal circuit (Figure 2.9): the dialyzer cartridge and the dialysate fluid. Figure 2.10 describes the construction of a dialyzer. It consists of a proper

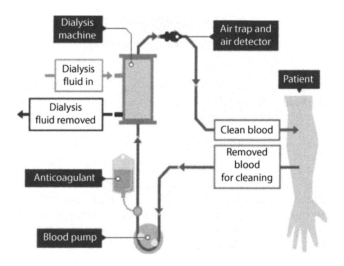

Figure 2.9   Hemodialysis circuit to purify blood.

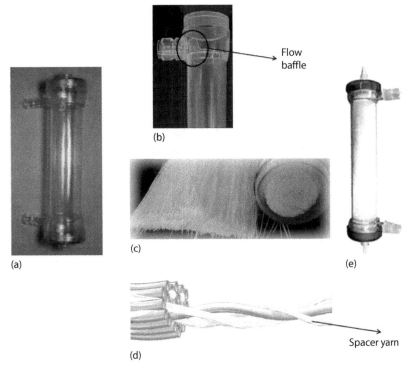

Figure 2.10   Hemodialyzer design (a) empty housing, (b) flow baffle, (c) hollow fiber membranes packed into the housing, (d) spacer yarns, and (e) complete hemodialyzer cartridge.

housing (Figure 2.10a), which should cater to the requirements of minimum transportation costs, storage costs, and minimum blood volume consumption during the dialysis treatment. Earlier, use of polycarbonate housing was widespread but now polypropylene housing has become popular, as they can be disposed of in an environmentally friendly manner.[25] As illustrated in Figure 2.9, the housing has two inlet and two outlet ports, for blood and dialysate fluids, respectively. The dialysate inlet port has a flow baffle (Figure 2.10b) that prevents direct impingement of dialysate fluid onto fibers, thereby reducing chances of rupture. The fibers, with clinical specifications as described in Figure 2.8, are bundled together and packed inside the housing. However, a couple of important points should be noted here. The first is the presence of spacer yarns as in Figure 2.10c and d. These are present to improve dialysate flow distribution, as well as maintain a cross flow pattern rather than parallel flow across fibers. This enhances the mass transport rate of uremic toxins.[26,27] Another technology in use to enhance mass transport is the use of undulated or crimped fibers.[28] The second point to be noted is the kind of potting material used for packing the fiber bundle (including spacer yarns). Polyurethane (PU) resin has classically been in use as a potting material. However, when

25

dialyzers are exposed to sterilization techniques like β- and γ-radiation, PU forms 4,4'-methylene dianiline, a fission product, which is a carcinogenic substance.[29] This also requires a compact design for the housing and the fiber bundle since it leads to less usage and exposure to PU. Hospal arylane series have reportedly used special treatment procedures to avoid complement activation during exposure to PU.[28] All these technologies integrated together give rise to a complete hemodialyzer design as depicted in Figure 2.10e.

In this context, a discussion about dialysate fluid is also necessary. It is important to note that dialysate affects the cardiovascular stability during the dialysis session. In fact, dialysate composition depends on various factors, including flow rate of blood and dialysate, treatment time, and even the choice of membrane. Hence, it is important to understand the composition of a typical dialysate along with the significance of each of the constituents in greater detail. This is provided in Table 2.1.

2.3.3 Clearance

Clearance (C_L') is defined as the removal rate (mL/min) of uremic toxins in a single pass. Its general equation is

$$C_L' = \frac{Q_{Bi}C_{Bi} - Q_{B0}C_{B0}}{C_{Bi}} \qquad (2.3)$$

where

Q_{Bi} is the inlet blood flow
Q_{B0} is the outlet blood flow
C_{Bi} is the inlet blood concentration of the uremic toxin
C_{B0} is the outlet blood concentration of the uremic toxin

This is the generalized equation that accounts for UF, that is, loss of fluid to the dialysate side. In case of zero UF, that is, with the inlet and outlet blood flows being equal, Equation 2.3 reduces to[31]

$$C_L = \frac{C_{Bi} - C_{B0}}{C_{Bi}} \times Q_B \qquad (2.4)$$

where

C_{Bi} and C_{B0} are the inlet and outlet solute concentrations in blood, respectively
Q_B is the blood flow rate

PROBLEM 2.1
Find the clearance of a typical hemodialyzer where $C_{Bi} = 600$ mg/L, $C_{B0} = 300$ mg/L, and blood flow rate $Q_B = 300$ mL/min.

Solution: Clearance is given by $C_L = \dfrac{C_{Bi} - C_{B0}}{C_{Bi}} \times Q_B$

or, $C_L = \dfrac{600 - 300}{600} \times 300 = 150$ mL/min.

Table 2.1 Typical Dialysate Composition and Significance of Components

Sl. No.	Component	Concentration in Dialysate Fluid	Significance
1.	Sodium	137 mEq/L	High concentrations lead to hypertension and pulmonary edema. In healthy patients, sodium balance is zero and this is also a desirable outcome of dialysis sessions. Various mathematical models have been proposed to model sodium transport during dialysis sessions.
2.	Chloride	105 mEq/L	Metabolic alkalosis may occur during dialysis, due to bicarbonate imbalance in the fluid, leading to nausea and headache. This is prevented by using high chloride concentration in dialysate fluid.
3.	Calcium	3.0 mEq/L	Dialysate consists of a protein-bound and a non-protein-bound fraction. Kidney malfunction leads to decreased ionized calcium levels in the body. Ultrafiltered calcium during dialysis is in ionized form and this form is responsible for blood coagulation, normal cardiac and skeletal muscle contraction, and nerve function.
4.	Acetate	4.0 mEq/L	Bicarbonate has steadily replaced its use in dialysate fluids. It was reported to cause myocardial contractility.
5.	Potassium	2.0 mEq/L	Ninety-eight percent of it is present in intracellular volume, and only 2% is present in plasma. This balance is maintained by the sodium–potassium pump, which lets three sodium ions out of cells for every two potassium ions into it. To ensure adequacy of potassium transport, removal during dialysis should equate the amount loaded.

(Continued)

Table 2.1 (*Continued*)   Typical Dialysate Composition and Significance of Components

Sl. No.	Component	Concentration in Dialysate Fluid	Significance
6.	Bicarbonate	33.0 mEq/L	Maintaining stable pH of blood. Acid production leads to hydrogen production, which is neutralized by bicarbonate, leading to formation of carbon dioxide and water. Kidney function deterioration leads to acid accumulation and a drop in the bicarbonate level in the body.
7.	Magnesium	0.75 mEq/L	It is absorbed in the small intestine and excreted by the kidneys to maintain stable levels in the human body. Kidney malfunction leads to a buildup of magnesium levels leading to altered nerve conduction velocity and increased pruritus.
8.	Dextrose	200 mg/dL	To enhance ultrafiltration by altering osmotic gradient.

Source: Locatelli, F. et al., *Nephrol. Dial. Transplant.*, 19, 785, 2004.

2.3.4 Ultrafiltration Coefficient (K_{UF})

The ultrafiltration coefficient (K_{UF}) is expressed in mL/h mmHg and is an indicator of the hydraulic permeability of membrane. It is evaluated in vitro using bovine blood, which is ultrafiltered at various TMPs. When TMP and the ultrafiltration flux are plotted, it becomes evident that they enjoy a linear relationship at low TMP and negligible variation at higher TMP.[32] The slope of this curve yields K_{UF}, which indicates the blood plasma filtration rate in dialysis.

2.3.5 Volumetric Mass Transfer Coefficient (K_oA)

K_oA is used as an indicator to classify different types of dialyzers. It is the product of the mass transfer coefficient K_o and the area of membrane, A[33]:

$$K_oA = \frac{1}{\dfrac{1}{Q_B} - \dfrac{1}{Q_D}} \times \ln\left(\frac{1 - \dfrac{C_L}{Q_D}}{1 - \dfrac{C_L}{Q_B}}\right) \qquad (2.5)$$

where

Q_B is the blood flow rate

Q_D is the dialysate flow rate

C_L is the clearance of the solute

An important point to be noted here is that the clearance used in Equation 2.5 can also be substituted as per Equation 2.3, if the blood flow in and blood flow out are different, in other words there is UF across the membrane.

PROBLEM 2.2

Given clearance, $C_L = 200$ mL/min, $Q_D = 500$ mL/min, and $Q_B = 300$ mL/min, calculate the volumetric mass transfer coefficient of a dialyzer.

Solution: $K_oA = \dfrac{1}{\dfrac{1}{Q_B} - \dfrac{1}{Q_D}} \times \ln\left(\dfrac{1 - \dfrac{C_L}{Q_D}}{1 - \dfrac{C_L}{Q_B}}\right)$

Hence, $K_oA = \dfrac{1}{\dfrac{1}{300} - \dfrac{1}{500}} \times \ln\left(\dfrac{1 - \dfrac{200}{500}}{1 - \dfrac{200}{300}}\right) = 750 \times 0.58 = 440.$

2.3.6 Dialysis Adequacy (*Kt/V*)

K indicates clearance (mL/min), t is the time of dialysis (minutes), and V is the volume of water in the patient's body (liters). For a normal adult patient of 60–70 kg body mass, a Kt/V value of 1.2 is considered to be adequate for a dialysis session.[34]

PROBLEM 2.3

Calculate the *Kt/V* ratio of a man weighing 70 kg using a dialyzer with clearance of 200 mL/min for 3 hours of dialysis procedure.

Solution: The man weighing 70 kg has a water content of $70 \times 0.6 = 42$ L $= 42,000$ mL. If the clearance rate of the dialyzer is 200 mL/min and the duration of dialysis is 3 hours or 180 minutes, then Kt/V is 0.85. This implies that the dialysis session is not adequate. To make it adequate, one may either use a dialyzer with higher clearance or expose the patient to a longer dialysis duration. However, the former, and not the latter, is desirable.

2.3.7 Sieving Coefficient

The movement of larger solutes is depicted by SC, or sieving coefficient, and is defined as[35]

$$SC = \frac{C_f}{C_b} \qquad (2.6)$$

where
 C_f is the solute concentration in the dialysate
 C_b is its feed concentration

PROBLEM 2.4
Calculate the sieving coefficient of a dialyzer where solute concentration in dialysate is 200 mg/L and that in feed is 400 mg/L.

Solution: The sieving coefficient is $SC = \dfrac{C_f}{C_b} = \dfrac{200}{400} = 0.5.$

Note: If the sieving coefficient is 0, it means that the solute does not permeate through and if it is 1 then it is a freely flowing solute whose concentration on both sides is equal. A typical example is sodium.

2.4 Hemodialysis Transport Mechanisms

In the context of hemodialysis, it is also important to understand the various transport processes involved and the different types of hemodialyzers available. The basic information needed would be the blood water flow of a patient. This is because water (of all the components of the blood) is ultrafiltered across the membrane carrying toxins with it, and hence this factor affects the transport rate. Thus, fixing blood flow alone would never help in realizing the real transport rates; hence, blood water flow has to be quantified to indicate the real transport. In this context, the dialyzer blood flow (Q_B) and blood water flow (Q_{Bw}) are defined as[35]

$$Q_{Bw} = Q_B \left[0.72\gamma(hct) + 0.93(1 - hct) \right] \qquad (2.7)$$

where
 hct is the fractional red cell volume, also known as hematocrit, of the blood
 γ is the fraction of red cell volume available for dialysis

The typical value of γ is 1.11 for urea and 0.5 for creatinine.[35] The water fraction is 0.93 in plasma and 0.72 in red cells.[35] Equation 2.6 also implies that reducing Q_B leads to a reduction in Q_{Bw}, and hence it is detrimental for solute clearance. This is the reason why blood flow rates in dialysis are kept at 200–300 mL/min.

PROBLEM 2.5
Given blood flow rate $Q_B = 300$ mL/min, find out the blood water flow rate of an adult male. Assume standard values of parameters and the solute to be urea.

Solution: For an adult male, fractional red cell volume is 0.45 and for women it is 0.4. Hence, $Q_{Bw} = Q_B[0.72\gamma(hct) + 0.93(1 - hct)]$

or $Q_{Bw} = 300[0.72 \times 1.11 \times 0.45 + 0.93 \times 0.55]$

hence, $Q_{Bw} = 261$ mL/min.

As described before, hemodialysis occurs when solutes from the fiber side permeate through the fiber side to the dialysate side. This is possible due to mass transfer that is facilitated by either concentration gradient or pressure differential. Another type of transport mechanism is the combination of both diffusion and convection. All three mechanisms are depicted in Figure 2.10. Two sections are divided by a semipermeable membrane that allows selective transport of species. Now, species transport from one section would be facilitated through the membrane into the adjacent section due to the concentration gradient existing between them. This follows from Fick's law of diffusion:

$$J = -D\frac{\partial C}{\partial x} \tag{2.8}$$

where
  J is the flux of solute
  D is the diffusivity of the solute
  $\dfrac{\partial C}{\partial x}$ is the concentration gradient

It is obvious from Equation 2.8 that the solute flux occurs in the direction of decreasing concentration, that is, the flux is from a higher to a lower concentration. This is an inherently slow process and is illustrated in Figure 2.11a. The other process (Figure 2.11b) is facilitated by applying pressure leading to the removal of solutes by convective transport. The last mechanism involves the application of both pressure and concentration gradients for solute transfer (Figure 2.11c). The first method is the classical and conventional hemodialysis mode of operation. The second mechanism is HF, and the third mode is HDF.[36]

In this regard, it is important to understand why and how these modes of dialysis developed over the years. This was due to not only deeper understanding about kidney failure and its ramifications, but also the development of material science and membrane engineering. One of the biggest challenges in dialysis therapy is the development of biocompatible membranes. This is discussed in the next section.

2.5 Hemodialysis Membranes: Material Development
2.5.1 Cellulose Acetate
As discussed previously, initially, dialysis membranes were made of cellophane or some naturally occurring materials. At this point, it is important

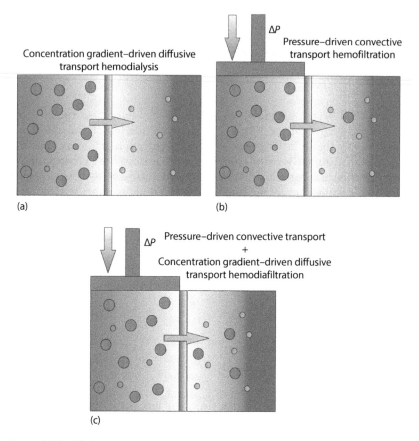

Figure 2.11   Various transport mechanisms involved in dialysis: (a) conventional dialysis, (b) hemofiltration, and (c) hemodiafiltration.

to introduce a debate that has been raging among nephrologists over the past three decades. This is regarding the middle molecule removal capacity (MMRC), which led to the "one meter square" hypothesis. This was first proposed by Babb et al.[37] in 1971. It was hypothesized that neuropathy (i.e., nerve disease which might be caused by dialysis) occurred due to an accumulation of middle molecules (MMs) that were not removed adequately during dialysis. In order to predict the removal of MMs, the "one square meter" hypothesis served as an index. It is the product of the number of hours of dialysis with the membrane surface area, and it was assumed that neither the dialysate flow rate nor blood flow rates were the influencing factors in removal of MMs. Rather both dialysis duration and higher surface area were the governing factors. The molecules responsible for occurrence of neuropathy in patients were the MMs like beta2 microglobulin (B2M). Hence, the removal of B2M was supposed to be dependent on the membrane surface area and hours of dialysis therapy and not on any other factors. Although the process was flawed to some extent since membrane permeability and pore sizes were not taken into account, it opened up new avenues of probing into dialysis therapy. Removal of MMs was also related to reduced cases of dialysis-related amyloidosis (DRA) occurring in dialysis patients. The crux of the matter is that besides removing urea and creatinine, the selected dialysis membrane must also remove MMs. As discussed in Chapter 1, it is imperative that the membrane starts mimicking the functions of the kidney. Cellulose acetate (CA) membranes were the first dialysis membranes. However, they inherently had very small pores that did not allow the removal of MMs, leading to increased cases of DRA. Moreover, CA membranes started exhibiting complement activation. Complements are groups of proteins responsible for the body's response toward antigens. These are activated in the presence of functional groups of CA. Apart from complement activation, leukopenia, hypoxia, and thrombocytopenia also occurred during dialysis sessions with CA membranes. Several variants of the cellulose family with commercial names of Cuprophan®, Bioflux®, and Cuprammonium Rayon were also employed. All had poor biocompatibility. To improve biocompatibility, it was decided that the acetate groups be replaced with acetyl, and hence cellulose diacetate and triacetate membranes came into use. These were more hydrophilic and were thus able to remove MMs better. These modified cellulosic family variants were used to fabricate commercial dialyzers with the trade names of Hemophan®, Synthetically Modified Cellulose (SMC)®, and Excebrane®. However, with this, another challenge arose and it was the thickness of the hollow-fiber membranes spun from the cellulose family. These dialyzers had thin walls (5–20 μm) because of which pressure differential could not be applied across them, and hence limited HF or HDF modes of dialysis were possible. This prompted the search for better membrane materials that (1) were more biocompatible, (2) had larger pores for removal of MM, and (3) were thick-walled to allow HF or HDF.

2.5.2 Polyacrylonitrile

Chenoweth et al.[38,39] demonstrated that polyacrylonitrile (PAN) membranes were better in preventing complement activation. This was because the complement activation mechanism released inflammatory proteins into the bloodstream, which initiated antibody reactions. However, PAN membranes adsorbed these proteins and prevented inflammatory response. This was a unique example where adsorption, an otherwise baneful mechanism in case of membrane separations, proved to be extremely useful. PAN membranes also adsorbed B2M, which was in the order of 400–600 mg/week.[39,40] The first PAN hollow-fiber membrane (commercial name AN69) was produced by *Rhone Poulenc* in France. It showed marked improvement over its predecessors, the CA family. However, PAN too had its limitations. It initiates severe anaphylactic reactions, which inspired the search for better membranes.

2.5.3 Polysulfone

The polysulfone (PSf) family, with larger pore sizes, proved to be a huge improvement over the CA and PAN families. It had higher diffusive clearances of MM solutes.[40,41] Scheider and Streicher reported a high B2M SC of 0.79[42] using a Fresenius F60 membrane. They reported that it had a high clearance for molecular weights between 11,400 and 40,000 Da. In fact, in line with Chapter 1, this is very close to the performance of the glomerulus. This class of membranes has been referred to as high-performance membranes (HPMs). The Japanese Society of Dialysis Therapy (JSDT) recommends a loss of albumin (<3 g/session) as this enhances the clearance of B2M and inflammatory cytokines.[43] This also leads to shorter dialysis periods, which improves the quality of life of patients.[44] Another advantage that this class of membranes enjoys is the ease of sterilization.[45] Previously, γ-rays and ethylene oxide (ETO) were employed, but various complications posed problems in their usage. For example, the extreme heat generated during γ-irradiation altered the cytotoxicity of membranes.[39] The PSf family solved this issue by simply allowing steam sterilization. Due to its excellent thermal stability, steam purging at 121°C helped in deactivating biological components and enabling the use of dialyzers.[45] Enhanced biocompatibility, reduced complement activation, and higher clearances of MMs have helped in boosting the popularity of PSf membranes.

2.5.4 Other Membrane Material for Dialysis

Ethylene vinyl alcohol copolymer (EVAL) and poly methyl methacrylate (PMMA) are two materials that are now in use for hemodialysis membrane synthesis. Akira Saito,[37] in collaboration with Kuaray Company, synthesized the first EVAL membrane with the desired pore size. Two commercial products, namely, KF101 C-2 (hemofilter) and C-2B (hemodialyzer), were developed.[46-49] Patients treated with EVAL dialyzers had fewer complaints of joint pain and DRA.[50] PMMA was developed due to the same problems associated with high-flux dialysis treatment. PMMA has high hydrophobicity, due to which the polymer does not swell in water, which in turn leads to crystal formation post phase inversion.[51] A different route was adopted to solve the issue, and this involved the formation of a stereocomplex between isotactic PMMA and syndiotactic PMMA. This in turn led to thermoreversible sol–gel phase transition, which can be explained on the basis of hydrogen bonding between the two PMMA structures.[51] PSf derivatives, namely, polyethersulfone (PES) and polyester polymer alloy (PEPA) dialyzers, are also commercially available.

Various companies use different base polymers to spin dialysis grade membranes. Regenerated cellulose was quite popular initially. Membrana manufactured two products using this polymer with the commercial names Cuprophan and Bioflux. They have a small wall thickness, varying from 5 to 17 μm. The advantages of these membranes are low cost, good solute removal, and satisfactory performance based on Kt/v. Major disadvantages associated with these membranes are poor removal capacity of middle-molecular-weight solutes, inferior biocompatibility, lower ultrafiltration coefficient (K_{UF}), and difficulty in conducting convection mode of dialysis due to lower wall thickness. Bioflux membranes had a larger pore diameter (7.23 Å). These membranes had several advantages over Cuprophan in terms of higher K_{UF} and K_oA, satisfactory removal of middle-molecular-weight solutes, and high mechanical strength. The Asahi/Tejni/Teruned/Toyobo companies had a product based on regenerated cellulose known as Cuprammonium Rayon. These membranes have an asymmetric structure, with a wall thickness of 9–26 μm, with higher biocompatibility. They have a larger pore size due to the use of PES as an additive, facilitating the removal of middle-molecular-weight solutes and can be categorized as high-flux dialysis membranes but with inferior biocompatibility.

Akzo came out with a product, Hemophan, using modified cellulose. The membranes have a wall thickness of 5.2 μm. Positively charged tertiary amino groups were introduced in its structure. These membranes have better biocompatibility and good performance for the removal of small solutes. But they have poor performance with respect to the removal of middle-molecular-weight solutes, lower K_{UF}, and poor biocompatibility. Furthermore, they are difficult to operate in the convective mode of dialysis due to thin walls.

Akzo/Tejin used SMC by substituting hydroxyl groups with charged benzyl groups. These fibers have better distribution of hydrophilic and hydrophobic groups than Hemophan along with lower complement activation; however, they have lower K_{UF} than Hemophan. Terumo used modified CA with the commercial name Excebrane. These membranes consisted of an outer surface with a higher-porosity cuprammonium layer and an inner surface exposing a thin acrylic layer. The addition of oleic acid reduced platelet activation and thrombogenic response. The copolymer coated with vitamin E reduced free radicals, thereby enhancing antioxidative capability. The advantage of these membranes was that they could be sterilized using an autoclave; however, they have poor biocompatibility.

Several companies brought out products based on acetylated cellulose. Althin/Toyobo introduced the first CA-based membranes. These were the most hydrophilic in the CA family but were also the least biocompatible. Hospal/Althin brought out cellulose diacetate membranes. They were more biocompatible, had higher K_{UF}, and could be sterilized with γ-radiation and ETO. But these membranes were more hydrophobic than CA and hence were prone to protein fouling and could not be sterilized with steam. Akzo/Toyobo used cellulose triacetate with 14–30 μm wall thickness, having higher K_oA and K_{UF} for both small- and middle-molecular-weight solutes. The membranes were suitable for HDF under convective mode. They could be sterilized with γ-radiation and ETO but could not be sterilized with steam.

PAN-based membranes were introduced by Rhone Poulenc. These membranes had a higher average pore size, 29 Å, with a wall thickness of 19–55 μm. They have high K_oA and K_{UF} for both small- and middle-molecular-weight solutes and have excellent removal capacity for B2M. They were suitable for both hemodialysis and HDF. But there is the likelihood of anaphylactic reactions at pH below 7.4.

Toray introduced PMMA-based membranes that were hydrophobic and asymmetric, with a wall thickness of 20–40 μm. They were good for B2M removal, biocompatible, and suitable for both hemodialysis and HDF. But being hydrophobic, these membranes were prone to fouling.

Gambro prepared membranes using a blend of polyethylene glycol (PEG) (20 wt%) and polycarbonate (80 wt%) (PPC Gambrane) with 15–60 μm wall thickness, good biocompatibility, lower thrombogenicity; they could be sterilized by γ-radiation or ETO. But these membranes are available in flat sheets only, having lower flux and lower K_oA.

Polyamide-based asymmetric membranes were made by Gambro, with a wall thickness of 50–60 μm. These membranes have excellent biocompatibility, are suitable for HDF, are sterilizable by ETO, and have low protein binding.

Membranes were prepared using polymers belonging to the PSf family by several companies like Fresenius, Asahi, Toray, and so on. These membranes have great chemical and thermal stability, high flux, and outstanding biocompatibility with high K_{UF} and are sterilizable by all modes.

EVAL-based membranes having good biocompatibility and higher permeation for middle-molecular-weight solutes were also synthesized. But, they were low-flux membranes for standard hemodialysis.

2.6 Conclusion

Biocompatibility and appropriate pore size distribution are key factors in the synthesis of dialysis membranes. In order to synthesize the right membrane for dialysis, the first step is to ensure the synthesis of a proper material that would prove biocompatible for dialysis patients. The synthesized material would then be employed for membrane engineering. This chapter revealed a lot about the dialysis membrane materials in use and the state-of-the-art technology available, apart from a possible starting point for dialysis membrane material synthesis. In this context, an important science to understand is "biocompatibility," which is discussed in the following chapter.

References

1. Graham, T. 1861. Liquid diffusion applied to analysis. *Philos. Trans. R. Soc. Lond.* 151: 183–224.
2. Abel, J.J., Rowntree, L.C., and Turner, B.B. 1913. On the removal of diffusible substances from the circulating blood by means of dialysis. *Trans. Assoc. Am. Phys.* 28: 51–62.
3. Abel, J.J., Rowntree, L.G., and Turner, B.B. 1913–1914. On the removal of diffusable substances from the circulating blood of living animals by dialysis. *J. Pharmacol. Exp. Ther.* 5: 275–316.
4. Abel, J.J., Rowntree, L.G., and Turner, B.B. 1913–1914. Some constituents of the blood. *J. Pharmacol. Exp. Ther.* 5: 611–623.
5. Abel, J.J., Rowntree, L.G., and Turner, B.B. 1913–1914. Plasma removal with return of corpuscles (plasmaphaeresis). *J. Pharmacol. Exp. Ther.* 5: 625–641.
6. Haas, G. 1923. Dialysieren des strömenden Blutes am Lebenden (Dialysis of the circulating blood in vivo). *Klin. Wochenschr.* 2: 1888 (in German).
7. Haas, G. 1925. Versuche der Blutauswaschung am Lebenden mit Hilfe der Dialyse (Experiments on cleansing of blood in vivo by means of dialysis). *Klin. Wochenschr.* 4: 13 (in German).

8. Haas, G. 1927. Ueber Versuche mit Blutwaschung am Lebenden mit Hilfe der Dialyse (On experimental cleansing of blood in vivo with dialysis). *Naunyn. Schmiedebergs. Arch. Pharmacol.* 120: 371 (in German).

9. Benedum, J. and Haas, G. 1979. Pionier der Hiimodialyse (Georg Haas: Pioneer of haemodialysis). *Med. Hist.* 14: 196–217 (in German).

10. Haas, G. 1928. Ueber Blutwaschung (On cleansing of blood). *Klin. Wochenschr.* 7: 1356–1362 (in German).

11. Kolff, W.J. and Berk, HTh.J. 1943. De kunstmatige nier: een dialysator met groot oppervlak (The artificial kidney: A dialyser with large surface area). *Ned Tijdschr Geneeskd* 87: 1684 (in Dutch).

12. Alwall, N. 1947. On the artificial kidney: Apparatus for dialysis of blood in vivo. *Acta Med. Scand.* 128: 317–325.

13. Alwall, N. and Norviit, L. 1947. On the artificial kidney. II. The effectivity of the apparatus. *Acta Med. Scand. (Suppl).* 196: 250–258.

14. Murray, G., Delorme, E., and Thomas, N. 1947. Development of an artificial kidney. *Arch. Surg.* 55: 505–522.

15. Garrelts, B. 1956. A blood dialyzer for use in vivo. *Acta Med. Scand.* 155(2): 87–94.

16. Inouye, W.Y. and Engelberg, J. 1953. A simplified artificial dialyzer and ultrafilter. *Surg. Forum* 4: 438–442.

17. Kiil, F. 1960. (Amundsen B): Development of a parallel flow artificial kidney in plastics. *Acta Chir. Scand. (Suppl).* 253: 142–149.

18. Kiil, F. and Glover, J.F. Jr. 1962. Parallel flow plastic hemodialyzer as a membrane oxygenator. *Trans. Am. Soc. Artif. Intern. Organs* 8: 43–46.

19. Loeb, S. and Sourirajan, S. 1962. Sea water demineralization by means of an osmotic membrane. *Adv. Chem. Series* 38: 117.

20. Loeb, S. 1981. The Loeb-Sourirajan membrane: How it came about. In: *Synthetic Membranes*, Turbak, A. (ed.). ACS Symposium Series 153, Vol. I, ACS, Washington, DC.

21. Pingsheng, H. 2013. *Structure and Properties of Polymers*. Alpha Science International, Limited, Oxford, UK.

22. Zeman, L.J. and Zydney, A.L. 1996. *Microfiltration and Ultrafiltration*. Marcel Dekker, New York.

23. Fujita, H. 1990. *Polymer Solutions*. Elsevier, Amsterdam, the Netherlands.

24. Matsuura, T. 1993. *Synthetic Membranes and Membrane Separation Processes*. CRC Press, Florida.

25. Mandolfo, S., Malbertio, F., Imbasciati, E., Cogliatii, P., and Gauly, A. 2003. Impact of blood and dialysate flow and surface on performance of a new polysulfone hemodialysis diaysers. *Intern. J. Artif. Organs* 26: 113–120.

26. Unger, J.K., Lemke, A.-J., and Grosse-Siestrup, C. 2006. Thermography as potential real-time technique to assess changes in flow distribution in hemofiltration. *Kidney Int.* 69: 520–525.

27. Poh, C.K., Hardy, P.A., Liao, Z., Huang, Z., Clark, W.R., and Gao, D. 2003. Effect of spacer yarns on the dialysate flow distribution of hemodialyzers: A magnetic resonance imaging study. *ASAIO J.* 49: 440–448.

28. Peinemann, K.-V. and Nunes, S.P. 2010. *Membrane Technology*, Vol. 1: Membranes for Life Sciences. Wiley-VCH Verlag GmbH & Co. KGaA, Weinheim, Germany.

29. Wilski, H. 1987. The radiation induced degradation of polymers. *Radiat. Phys. Chem.* 29: 1–14.

30. Locatelli, F., Covic, A., Chazot, C., Leunissen, K., Luño, J., and Yaqoob, M. 2004. Optimal composition of the dialysate, with emphasis on its influence on blood pressure. *Nephrol. Dial. Transplant.* 19: 785–796.

31. Depner, T.A. 2005. Hemodialysis adequacy: Basic essentials and practical points for the nephrologist in training. *Hemodialysis Int.* 9: 241–254.
32. Henderson, L. 1996. Biophysics of ultrafiltration and hemofiltration. In: *Replacement of Renal Function by Dialysis,* 4th edn., Jacobs, C., Kjellstrand C., Koch, K., and Winchester J. (eds.). Kluwer Academic Publishers, Dordrecht, the Netherlands, pp. 114–145.
33. Yamashita, A.C. 2011. Mass transfer mechanisms in high-performance membrane dialyzers. *Contrib. Nephrol.* 173: 95–102.
34. Daugirdas, J.T. 1993. Second generation logarithmic estimates of single-pool variable volume Kt/V: An analysis of error. *J. Am. Soc. Nephrol.* 4(5): 1205–1213.
35. Depner, T. and Garred, L. 2004. Solute transport mechanisms in dialysis. In: *Replacement of Renal Function by Dialysis,* Jacobs, C., Kjellstrand C., Koch, K., and Winchester J. (eds.). Springer, Dordrecht, the Netherlands, pp. 73–94.
36. Sprenger, K.B.G., Kratz, W., Lewis, A.E., and Stadtmuller, U. 1983. Kinetic modeling of hemodialysis, hemofiltration, and hemodiafiltration. *Kidney Int.* 24(2): 143–151.
37. Babb, A.L., Popovich, R.P., Christopher, T.G., and Scribner, B.H. 1971. The genesis of square-metre-hour hypothesis. *Trans. Am. Soc. Artif. Intern. Organs* 17: 81–91.
38. Chenoweth, D.E., Cheung, A.K., and Ward, D.M. 1983. Anaphylatoxin formation during hemodialysis: Effects of different dialyzer membranes. *Kidney Int.* 24: 764–769.
39. Chenoweth, D.E., Cheung, A.K., Ward, D.M., and Henderson, L.W. 1983. Anaphylatoxin formation during hemodialysis: Comparison of new and used dialyzers. *Kidney Int.* 24: 770–774.
40. Goldman, M., Dhaene, M., and Vanherweghem, J.-L. 1986. Removal of β_2-microglobulin by adsorption on dialysis membranes. *Nephrol. Dial. Transplant.* 2: 576–577.
41. Zingraff, J., Beyne, P., Urena, M., Uzan, M., Nguyen Khoa, M., Descamps-Latscha, B., and Drueke, T. 1988. Influence of hemodialysis membranes on β_2-microglobulin kinetics: In vivo and in vitro studies. *Nephrol. Dial. Transplant.* 3: 284–290.
42. Rockel, A., Hertel, J., Fiegel, P., Abdelhamid, S., Panitz, N., and Walb, D. 1986. Permeability and secondary membrane formation of a high flux polysulfone hemofilter. *Kidney Int.* 30: 429–432.
43. Saito, A. 2011. Definition of high-performance membranes—From the clinical point of view. In: *High-Performance Membrane Dialyzers.* Contributions to Nephrology, Vol. 173, Saito, A., Kawanishi, H., Yamashita, A.C., and Mineshima, M. (eds.). Karger, Basel, Switzerland, pp. 1–10.
44. Kolff, W.J. and Watschinger, B. 1956. Further development of a coil kidney. Disposable artificial kidney. *J. Lab. Clin. Med.* 47: 969–977.
45. Bowry, S.K., Gatti, E., and Vienken, J. 2011. Contribution of polysulfone membranes to the success of convective dialysis therapies. *Contrib. Nephrol.* 173: 110–118.
46. MacLeod, A.M., Campbell, M.K., Cody, J.D., Daly, C., Grant, A., Khan, I., Rabindranath, K.S., Vale, L., and Wallace, S.A. 2005. Cellulose, modified cellulose and synthetic membranes in the haemodialysis of patients with end-stage renal disease (Review). *Cochrane Database of Systematic Reviews.*

47. Saito, A., Chung, T.G., Kanazawa, I., Oda, O., and Ohta, K. 1981. Clinical evaluation of protein permeating hemofilter and analysis of middle molecules in ultrafiltrate (in Japanese). *Jinkou-Zouki* 10: 907–911.

48. Saito, A., Chung, T.G., Kanazawa, I., Ogawa, H., and Ohta, K. 1982. Middle and large molecule removal of protein-permeating hemodiafiltration. *Proceedings of the Fourth International Symposium on Hemoperfusion and Artificial Organs*, Ankara, Turkish, Artificial Organs Society, Piskin, E. and Chang, T.M.S. (eds.), pp. 42–46.

49. Saito, A., Ogawa, H., Takagi, H., and Chung, T.G. 1983. Clinical effects of large molecule removal by protein-permeable hemofilters (in Japanese). *Jin To Touseki* 15: 767–772.

50. Saito, A., Naito, H., and Hirohata, M. 1984. Dialytic removal of middle molecules and low molecular weight proteins. In: *Progress in Artificial Organs*, Atsumi, K., Maekawa, M., and Ota, K., (eds.). ISAO Press, Cleveland, OH, pp. 413–416.

51. Sakai, Y. and Tanzawa, H. 1978. Poly(methyl methacrylate) membranes. *J. Appl. Polym. Sci.* 22: 1805–1815.

52. Twardowski, Z.J. 2008. History of hemodialyzers' designs. *Hemodial. Int.* 12(2): 173–210.

53. Agar, J., Schatell, D., and Witten, B. n.d. Kolff-Brigham dialysis machine: 1948. http://www.homedialysis.org/home-dialysis-basics/machines-and-supplies/dialysis-museum, retrieved November 24, 2016.

Biocompatibility and Biomaterials

The art of medicine consists in amusing the patient while nature cures the disease.

—**Voltaire**

3.1 Biocompatibility and Biomaterials: Definition

According to Williams (1987),[1] "a biomaterial is a nonviable material used in a medical device, intended to interact with biological systems" and "biocompatibility is the ability of a material to perform with an appropriate host response in a specific application."

Developing biomaterials is a dynamic field of research, with contributions from various disciplines ranging from chemistry and biology to chemical, mechanical, and electrical engineering. However, developing biomaterials requires an in-depth understanding of biocompatibility. Unfortunately, measuring biocompatibility is always challenging due to the lack of precise definitions. Moreover, it has to be uniquely defined for each application. For example, biocompatibility issues of a bone scaffold are entirely different from those of a hemodialysis membrane. Hence, application-specific definition renders different points of view of biocompatibility when dealing with soft tissue, hard tissue, and the cardiovascular systems, as well as for those that are placed in the extracorporeal circuit. However, even without a proper grasp of the concepts of biocompatibility, biomaterials have been invented and tried over subjects, intentionally or accidentally, for centuries. In fact, the example of "Kennewick Man," dating back 9000 years, is a classic example in this regard.[2] This is the case of the remains of a man unearthed from Kennewick, Washington, USA, whom archaeologists described as a tall, healthy, and active person, with the only blemish being a spear point embedded in his hip, which had healed in and reportedly did not impede any of his normal activities. This was an accidental implant of a foreign object into the human body, which seemingly did not alter the body's capacity due to its presence.[2] To understand how necessity became the mother of invention in the case of biomaterials, it is important to have a brief look into its history.

3.2 History of Biomaterials
3.2.1 Early Development
3.2.1.1 Dental Implants
As early as 600 AD, the Maya civilization invented teeth implants from seashells, seamlessly fitting into the existing teeth.[3] This is known as bone integration and was performed for centuries without proper knowledge of biomaterial science. However, this was planned and not accidental as in the case of the "Kennewick Man." Similarly, in Europe too, an iron dental implant was reported dating back to 200 AD.[4]

3.2.1.2 Sutures
Linen sutures were used by the early Egyptians and catgut was used during the Middle Ages in Europe.[2] Greek literature was the first to report the use of metallic sutures, and in 1816 Philip Physick of the University of Pennsylvania suggested the use of lead wire sutures.[2]

3.2.1.3 Artificial Heart
Aristotle termed the heart as "the most important organ of the human body."[2] It was in 1628 that William Harvey first described the heart as a pump. This prompted French physiologist Le Gallois to hypothesize that organs could be kept alive by pumping blood through them. Étienne-Jules Marey first described an artificial heart device but the first realizable design and patent came forth in the 1950s, which was realized through the efforts of Dr. Paul Winchell.

3.2.1.4 Contact Lenses
Leonardo Da Vinci first developed the concept of contact lens in 1508. Rene Descartes (1632) developed the idea of corneal contact lens and Sir John F. W. Herschel (1827) proposed that a glass lens could protect the eye. Adolf Fick (propounded the law of diffusion) first developed a glass contact lens. Polymethylmethacrylate (PMMA)-based polymeric contact lens was developed between 1936 and 1948.

3.2.2 Post–World War II: Surgeon Hero Era
In this regard, it is very important to understand the contribution of medical practitioners during World War II toward the field of biomaterial development. The war crippled resources as well as chances of collaborative learning between continents and countries. It also saw minimum government regulations being imposed and little or almost nil human protection from medical trials. Under such circumstances, it is easy to imagine that doctors enjoyed

unlimited freedom and at the same time were burdened with the enormity of fully assuming the responsibilities of a patient's life. Such juxtaposition of emotions often led the experts to take a "huge technological/professional leap"[2] to restore the patient's life or organ. This led to the coinage of the term "Surgeon Hero," which placed the medical practitioners as heroes in eyes of the common man, patients, and soldiers alike. This era witnessed the laying of the foundation of modern-day biomedical engineering and materials for biological applications.

3.2.2.1 Dental Implants
Around 1937, two attempts by Venable (Co–Cr–Mo alloy) and Strock (screw-type implant of vitallium) were reported for surgical dental implants. However, a significant advance was reported by Per Ingvar Branemark in 1952, which was rather fortuitous. He observed that a titanium cylinder screwed into a rabbit bone healed and integrated completely. Later Dr. Branemark coined the term "osseointegration" and now most dental implants are made of titanium and its alloys.

3.2.2.2 Intraocular Lenses
Sir Harold Ridley (1906–2001) investigated the eyes of aviators after the war. He found the accidental occurrence of shards in the eyes of pilots, which was attributed to explosions occurring in shattered canopies. These had not caused any damage to the eyes and, in fact, had healed in place without any irritation. The material was traced back and found to be PMMA.[2] Needless to say, he designed, fabricated, and implanted the first intraocular lens on November 29, 1949. Although he faced a lot of social challenges in the acceptance of his ingenuity, this "Surgeon Hero" approach led to the creation of a multibillion dollar industry later on.

3.2.2.3 Hip/Knee Prostheses
Materials ranging from cemented ivory balls to glass hemispheres were tried for hip replacement without much success. However, a Teflon acetabular cup in 1958 and later on in 1961 high-molecular-weight polyethylene were used by Dr. Charnley. This proved to be very successful, and it was further refined by introducing PMMA cements. The developments in hip prosthesis provided some cues for the knee joint genre, and surgeons Frank Gunston and John Insall in 1968–1972 came up with encouraging results.

3.2.2.4 Artificial Kidney
The subject of artificial kidney has already been discussed in Chapter 2.

3.2.2.5 Artificial Heart
Willem Kolff, a name already familiar in the field of artificial kidneys, also played a pioneering role in the development of the artificial heart. Thermosetting poly(vinyl chloride) was used to fabricate the Kolff heart and implanted in a dog in 1957.[2] However, Dr. Michael DeBakey implanted a left ventricular assist device in 1966, and a polyurethane-based complete artificial heart was implanted in 1969 by Dr. Denton Cooley.

3.2.2.6 Vascular Grafts and Stents

Damaged and diseased blood vessels are one of the most common challenges encountered by surgeons. Vitallium metal tubes were put to use to mend arterial defects by Blackmore in 1942.[2] A serendipitous observation by the surgeon Authur Voorhees (1922–1992) was that tissue growth around silk fibers was possible in vivo, and this inspired experimentation using a silk handkerchief and parachute fabric. Reports and papers were published demonstrating encouraging observations regarding the use of a porous fabric wrapped around a solid polyethylene tube.[2,5]

3.2.2.7 Pacemakers

During 1820–1880, it was already known that electricity could alter heartbeats, but it was not until 1949 that John Hopps (a Canadian electrical engineer) discovered that if a cooled-down heart stopped beating then it could be started with electrical shocks. He invented the vacuum tube pacemaker in 1950. During 1957–1958, the founder of Medtronic, Earl E. Bakken, developed the first wearable pacemaker. Wilson Greatbatch and cardiologist W. M. Chardack, in 1959, developed the first fully implantable wearable pacemaker.

3.2.2.8 Heart Valves

PMMA tubing and nylon ball formed the first implantable heart valve, tried out heroically and successfully by Charles Hufangel in 1952. Later on Gibbon (1953) and Albert Starr (1960) carried out evolutionary changes in the design. The first leaflet tissue heart valve was developed by Warren Hancock in 1969 and was later acquired by Johnson & Johnson in 1979.

3.3 Essential Properties of Biomaterials

It is not difficult to comprehend that in order to design the right kind of device or implant for biological processes, the exercise should be carried out based on some solid fundamental findings rather than random selection. In this regard, it is important to understand the essential subjects related to biomaterials science.[2]

3.3.1 Toxicology

This is the simplest and most basic property relevant for any biomaterial, that is, the biomaterial should not be toxic, unless otherwise designed specifically for such an operation. A simple example is any drug delivery system designed to destroy specific cells, like cancer cells. Toxicological testing for biomaterials has evolved into a complicated branch of science. Mainly, this deals with "leachables" or the substances that leach out of biomaterials as these possess the potential to exhibit toxicity toward the physiological functions and cells of the body.

3.3.2 Biocompatibility

Thus, it is evident that biocompatibility is one of the properties of a biomaterial. Unfortunately, as discussed previously, biocompatibility is based more on experience and observations and it has no concrete definition. Moreover, it often refers to performing a specific task.

3.3.3 Functional Tissue Structure and Pathobiology

These are important when a device is designed for in vivo application. Normal, abnormal cell growth, tissue interaction, as well as disease processes are studied under this.

3.3.4 Healing

Well-defined inflammatory response is studied in these cases, leading to the process of healing. In-depth understanding of how a foreign object alters inflammatory response forms the line of investigation.

3.3.5 Mechanical Requirements

Each type of biomaterial has a particular mechanical property requirement. Three categories of requirement would be mechanical performance, mechanical durability, and physical properties. Mechanical performance can be understood with the example of hip prostheses, which must be strong and rigid. Mechanical durability can be understood with the example of catheter, which has to function for 3 days, or a bone plate, which may have to function for 6 months or more. Physical properties can be understood by using a dialysis membrane. The structure and porosity of a membrane determine its capability to transport uremic toxins. Hence, engineering the materials with the necessary mechanical requirements for specific applications is a challenge.

3.3.6 Miscellaneous and Peripheral Requirements

Apart from the properties mentioned earlier, the development of biomaterials and practical implementation of the same require industrial involvement. This is a crucial aspect in the world of biomaterials, since an active interface between research and development of products and taking them to the market is very important in this field. Else, the right solution would never reach the patients. However, in this regard, ethics and regulations play an important role. Animal model testing of biomaterials, their justifications, minimizing investigator, and/or company bias toward biomaterial development, and so on, have to be delved into. Moreover, the investments for introducing biomaterial are often huge, which should also be taken into consideration.

3.4 Biocompatibility in Case of Dialysis Membranes

From the discussions earlier, it has to be understood that the synthesis of a dialysis membrane is basically the synthesis of a biomaterial. However, it has to undergo testing in order to qualify as a solution. For this, it is important to understand cytocompatibility and hemocompatibility assays that a particular synthesized membrane has to undergo.

3.4.1 Cytocompatibility

In case of hemodialysis membranes, cytocompatibility tests commonly encompass cell metabolic tests (or assays), cell proliferation assay, protein adsorption, and cytotoxicity tests. Each of these is discussed as follows.

3.4.1.1 Cell Metabolic and Cell Proliferation Assays

It is customary to estimate cell metabolism, in the case of hemodialysis applications, using MTT, which is a tetrazolium dye (3-(4,5-dimethylthiazol-2-yl)-2,5-diphenyltetrazolium bromide). Cell metabolism produces oxidoreductase enzymes, which reduce the MTT dye to formazan (Figure 3.1). MTT is yellow in color and turns to purple on reduction. This color change occurs in living cells. These then undergo a solubilization process involving dissolution in solvents like dimethyl sulfoxide or acidified ethanol solution. The dissolution results in forming a colored solution whose absorbance can be measured using a spectrophotometer. Higher the degree of absorbance more is the metabolic activity and vice versa.

Cell proliferation is estimated through quantification of deoxyribonucleic acid (DNA) for evaluating the cell number.[6] The method is accurate and sensitive to an appreciable degree and has gained popularity for evaluating the differentiation and proliferation of cells. It is based on enzymatic digestion followed by mechanical or chemical lysis.[7,8] The DNA is then homogenously dispersed in the assayed extraction solution, bound to a fluorescent dye, and the intensity of color developed is measured fluorometrically.[9]

3.4.1.2 Oxidative Stress

In vitro cytotoxicity assays prove helpful in multiple ways. They help in reducing animal experimentation, by analyzing the toxicity of developed drugs, and evaluate their safety, carcinogenicity, as well as chronic toxicity. Hence, in vitro cytotoxicity of developed materials can be tested on cell lines to investigate cell lysis, necrosis, or loss of membrane integrity. The oxidative stress developed is determined by measuring the reactive oxygen species (ROS) grown inside the cells. ROS are chemically reactive molecules containing oxygen, for example, peroxides, superoxides, hydroxyl radicals, and singlet

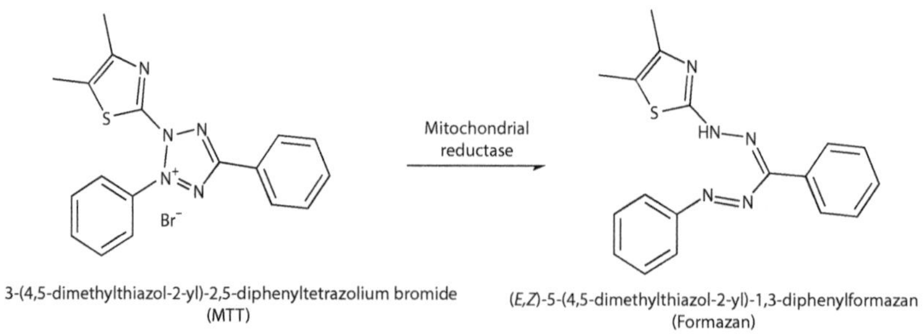

3-(4,5-dimethylthiazol-2-yl)-2,5-diphenyltetrazolium bromide (MTT)

(E,Z)-5-(4,5-dimethylthiazol-2-yl)-1,3-diphenylformazan (Formazan)

Figure 3.1   Reduction of MTT to formazan.

46

oxygen. ROS is the by-product of normal metabolism of oxygen, which can increase several folds due to external stimuli such as heat, chemicals, smoke, and tobacco. Such external stimuli can strip water of an electron that becomes a highly active species, thereby producing hydroxyl radical ($\cdot$OH), hydrogen peroxide (H_2O_2), superoxide radical ($\cdot O_2-$), and ultimately oxygen (O_2) in a step-by-step sequence.[10] The hydroxyl radical can strip molecules of electrons initiating a chain reaction, whereas hydrogen peroxide is very damaging to DNA. The ROS activity can be measured by the dichloro-dihydrofuran-fluorescein diacetate (DCFH-DA) assay.[11] It is a fluorogenic dye measuring hydroxyl, peroxyl, and other ROS activity within the cell. It diffuses into the cell and gets deacetylated by cellular esterases to a nonfluorescent compound, which is later oxidized by ROS into 2',7'-dichlorofluorescein (DCF), a highly fluorescent species.[11] Thus, measuring the fluorescence directly provides a measure of the ROS activity.

3.4.1.3 Protein Content
Protein content is estimated via the bicinchoninic acid (BCA) assay.[12] When placed in alkaline solution containing Cu^{2+} ions proteins, it forms a colored complex between the peptide bonds and copper atoms. This direct technique was, however, inefficient to detect low concentrations and hence the Lowry assay was developed, where the Folin–Ciocalteu reagent was used to enhance color. BCA is a substitute for the Folin–Ciocalteu reagent forming a 2:1 complex with protein resulting in a stable and highly colored complex detectable at 562 nm.[12]

3.4.2 Hemocompatibility
Hemocompatibility tests are vital for applications like hemodialysis. Standard hemocompatibility tests carried out to investigate the viability of a membrane synthesized for hemodialysis applications are hemolysis, blood cell aggregation, platelet adhesion, and thrombus formation.

3.4.2.1 Hemolysis
This is the lysis or rupturing of red blood cells. Generally, there can be several reasons behind hemolysis, for example, medical conditions, autoimmune disorders, genetic disorders, or blood with low solute concentration. It can also happen outside the body, due to mechanical damage to RBCs caused during surgery or bacterial culture or even exposure to surfaces non-conducive to blood, as in the case of dialysis membranes.

3.4.2.2 Blood Cell Aggregation
Erythrocytes or red blood cells can aggregate under abnormal conditions, where the individual cells "stick" together and give the impression of a viscous fluid behaving like glue. The aggregation forms in a special way, forming a rouleaux.[13] Rouleaux are stacks of erythrocytes, and stacks can form due to the unique discoid shape of the cells. It occurs in normal blood under low-flow conditions or at stasis. The underlining mechanism for rouleaux formation is depletion of high-molecular-weight proteins, like fibrinogen, also known as

chemiosmotic hypothesis for aggregation. Aggregation can occur under various conditions such as infection, inflammatory and connective tissue disorders, and cancers.

3.4.2.3 Platelet Adhesion and Thrombus Formation

Platelet adhesion is an essential function in response to vascular injury. Platelets adhere to particulate matter in the bloodstream, bacteria, and microorganisms, as well as artificial surfaces of prosthetic devices, bio-incompatible dialysis membranes, and so on. In fact, platelet adhesion is the first step toward thrombus formation, arresting hemorrhage and permitting wound healing. Platelet aggregation can also be triggered by blood flow or high shear flow conditions. In fact, the Von Willebrand (vWF) factor is used to understand both the phenomena. vWF is a blood glycoprotein that is involved in homeostasis. It has been reported that vWF binds to collagen, tethering platelets to the collagen surface and thus contributing to thrombus formation as well.[14]

3.5 Essential Equipment for Analyzing Biocompatibility

In the discussions earlier, it is evident that several equipment are needed to analyze and evaluate the biological responses of surfaces. Few of the relevant equipment are discussed in the succeeding chapter, but two basic equipment that need a mention here before going any further are the spectrophotometer and fluorescence microscope.

3.5.1 Spectrophotometer

It works on the simple principle of reflection and transmission properties of a material or a solution. When a monochromatic light passes through a solution (concentration c), there exists a relationship between light that is transmitted through the solution and the concentration of the solute in the solution, expressed as[15]

$$I = I_0 10^{-kcl} \tag{3.1}$$

where
  I is the intensity of transmitted light in solution
  I_0 is the intensity of light when the solvent is pure
  k is a constant
  l is the light path that is a constant for a test cell

Equation 3.1 is also known as Beer's law. Figure 3.2a depicts the working principle, and Figure 3.2b and c illustrates a typical spectrophotometer. Light from a source lamp passes through a monochromator (Figure 3.2a). It gets diffracted into the constituent wavelengths and narrow bandwidths of the diffracted spectrum pass through the adjustable slit and the test sample, and the photon density of the transmitted sample is measured using a photodiode. The transmittance of the sample is compared with the reference sample and

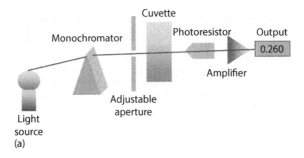

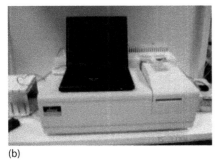

(b)   (c)

Figure 3.2 (a) Principle of spectrophotometer operation, (b) typical spectrophotometer, and (c) chamber for placing cuvette.

determines absorbance values, from which concentration can be calculated. Usually, spectrophotometers are of two classes, single beam and double beam. Single beam measures relative intensity of light before and after the sample is inserted, whereas in double beam, the measurement is carried out simultaneously. A typical spectrophotometer is depicted in Figure 3.2b. Figure 3.2c shows the chamber of the cuvette of a double-beam spectrophotometer. The two cuvettes (reference and sample) are placed in the sockets, and the beam passes through both of them for analysis.

3.5.2 Fluorescent Microscope

The fluorescent microscope is a class of optical microscope but it uses the principle of fluorescence and phosphorescence instead of relying only on reflection and adsorption. The specimen is illuminated by a light of known wavelength and the fluorophores absorb the light, emitting a light of different wavelength. The most conventional way to prepare a fluorescent sample is by using a fluorescent stain, or if the sample itself has fluorescent properties, that is, autofluorescence. Typical components of a fluorescent microscope are a light source (xenon lamp or mercury lamp), an excitation filter, a dichroic mirror, and an emission filter (Figure 3.3a). The filters and dichroic are chosen one at a time in order to match the spectral excitation and emission characteristics of the fluorophore. If there are multicolored fluorophores present in the sample, then

49

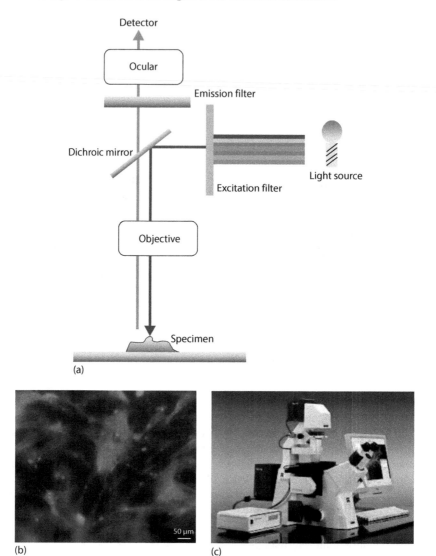

Figure 3.3 (a) Working principle of a fluorescent microscope, (b) typical fluorescent microscope, and (c) fluorescent image of NIH3T3 cells taken in a fluorescent microscope.

multiple images of the single specimen (each of a specific color) are combined to generate the collective image of the specimen. A typical fluorescence microscope is illustrated in Figure 3.3b, and a representative fluorescent image if NIH3T3 cells are presented in Figure 3.3c. The nucleus is stained with DAPI (a blue-colored dye).

3.6 Conclusion

With the foundation laid on proper understanding of biomaterials and its development and basics of biocompatibility, it is time to synthesize a proper biocompatible material suitable for dialysis. It is clear from the discussions in the chapters that in order to synthesize the correct dialysis membrane, it is important to choose the right base polymer. Second, the membrane should be cytocompatible and hemocompatible. Third, the membrane should have the right pore size and physical characteristics in order to permeate uremic toxins. Keeping all these in mind, efforts are undertaken in the next chapter to synthesize and determine the perfect composition to synthesize a dialysis membrane.

References

1. Williams, D.F. 1987. Definitions in biomaterials. *Proceedings of a Consensus Conference of the European Society for Biomaterials*, Chester, U.K., March 3–5, 1986, Vol. 4, Elsevier, New York.
2. Ratner, B.D., Hoffman, A.S., Schoen, F.J., and Lemons, J.E. 2004. *Biomaterials Science: An Introduction to Materials in Medicine*. Academic Press, Cambridge, MA.
3. Bobbio, A. 1972. The first endosseous alloplastic implant in the history of man. *Bull. Hist. Dent.* 20: 1–6.
4. Crubezy, E., Murail, P., Girard, L., and Bernadou, J.-P. 1998. False teeth of the Roman world. *Nature* 391: 29.
5. Egdahl, R.H., Hume, D.M., and Schlang, H.A. 1954. Plastic venous prostheses. *Surg. Forum* 5: 235–241.
6. Piccinini, E., Sadr, N., and Martin, I. 2010. Ceramic materials lead to underestimated DNA quantifications: A method for reliable measurements. *Eur. Cell Mater.* 20: 38–44.
7. Cui, L., Liu, B., Liu, G., Zhang, W., Cen, L., Sun, J., Yin, S., Liu, W., and Cao, Y. 2007. Repair of cranial bone defects with adipose derived stem cells and coral scaffold in a canine model. *Biomaterials* 28: 5477–5486.
8. Du, D., Furukawa, K., and Ushida, T. 2008. Oscillatory perfusion seeding and culturing of osteoblast-like cells on porous beta-tricalcium phosphate scaffolds. *J. Biomed. Mater. Res. Part A* 86A: 796–803.
9. Roy, A., Dadhich, P., Dhara, S., and De, S. 2015. In vitro cytocompatibility and blood compatibility of polysulfone blend, surface-modified polysulfone and polyacrylonitrile membranes for hemodialysis. *RSC Adv.* 5(10): 7023–7034.
10. Turrens, J.F. 2003. Mitochondrial formation of reactive oxygen species. *J. Physiol.* 552(Pt 2): 335–344.
11. Aranda, A., Sequedo, L., Tolosa, L., Quintas, G., Burello, E., Castell, J.V., and Gombau, L. 2013. Dichloro-dihydro-fluorescein diacetate (DCFH-DA) assay: A quantitative method for oxidative stress assessment of nanoparticle-treated cells. *Toxicol. In Vitro* 27(2): 954–963.
12. Smith, P., Krohn, R.I., Hermanson, G.T., Mallia, A.K., Gartner, F.H., Provenzano, M., Fujimoto, E.K., Goeke, N.M., Olson, B.J., and Klenk, D.C. 1985. Measurement of protein using bicinchoninic acid. *Anal. Biochem.* 150(1): 76–85.

13. Stoltz, J.F., Gaillard, S., Paulus, F., Henri, O. and Dixneuf, P., 1983. Experimental approach to rouleau formation. Comparison of three methods. *Biorheology. Supplement: the official journal of the International Society of Biorheology,* 1, 221–226.
14. Tomokiyo, K., Kamikubo, Y., Hanada, T., Araki, T., Nakatomi, Y., Ogata, Y., Jung, S.M., Nakagaki, T., and Moroi, M. 2005. Von Willebrand factor accelerates platelet adhesion and thrombus formation on a collagen surface in platelet-reduced blood under flow conditions. *Blood* 105(3): 1078–1084.
15. Schwedt, G. 1997. *The Essential Guide to Analytical Chemistry* (Brooks Haderlie, trans.). Wiley, Chichester, U.K., pp. 16–17 (Original work published 1943).

4

Selection of Material for Dialysis Membrane

Natural selection eliminates and maybe maintains, but it doesn't create.

—**Lynn Margulis**

4.1 Introduction

In view of the previous chapters, it is necessary to select the right material for hemodialysis application such that it is versatile in terms of hemocompatibility, cytocompatibility, as well as desired transport capability of uremic toxins. Moreover, the previous chapters also helped in understanding the starting point of the venture. It is clear that cellulose acetate, or its derivatives, would never be an intelligent choice for various biophysical problems. A more rationale choice might be polyacrylonitrile (PAN), but polysulfone (PSf) is the preferred choice for the base material for a likely dialysis membrane. The basic objective of this chapter is to identify the polymeric composition for synthesis of a hemodialysis membrane. Hence, the starting point in this endeavor is the synthesis of a typical, flat-sheet high-efficiency (6 kDa) membrane. Three different routes have been tried for this purpose. First, a polymer blend membrane was synthesized with PSf as the base material, mixed with polyvinylpyrrolidone (PVP) and polyethylene glycol (PEG). Second, a PSf–PVP blend membrane was synthesized, which was surface-treated with trimesoyl chloride (TMC) and m-phenylene diamine (MPD) to yield 6 kDa molecular weight cutoff (MWCO). Finally, a PAN (homopolymer) membrane was synthesized based on work by Roy and De.[1] Hereon, the blend membrane will be referred to as S1, the surface-modified membrane will be referred to as S2, and the PAN membrane will be referred to as S3. The three membranes were characterized comprehensively in terms of their hydraulic permeability,

MWCO, hydrophilicity, porosity, membrane morphology, surface charge, and mechanical strength. Once the physical characteristics of the three membranes were understood, they were subjected to comprehensive cytocompatibility and hemocompatibility analysis. The three membranes were finally subjected to tests for uremic transport. On the basis of transport and biological response, the most suitable solution was selected to proceed toward the challenging task of process design for spin dialysis membranes.

4.2 Materials and Methods

In this section, the physical characterization of membranes and the associated materials and methods have been described. The procedure followed in this section remains the same for the succeeding chapters, unless otherwise mentioned.

4.2.1 Polymer Solution Synthesis
4.2.1.1 S1 Solution
Membrane synthesis, as discussed before, requires synthesis of a lacquer or polymeric solution; 18 wt% PSf (average molecular weight 22,400 Da, supplied by M/s. Solvay Chemicals, Mumbai, India), 1 wt% PVP (molecular weight 40,000 Da, supplied by M/s. Sigma Aldrich, Missouri, USA), and 3 wt% PEG (supplied by M/s. S R Ltd., Mumbai, India) were dissolved in dimethylformamide (DMF) (supplied by M/s. Merck [India] Mumbai Ltd.). Dissolution was carried out in a flat bottom flask and slow stirring with a magnetic stirrer at a constant temperature of 60°C for 10 hours. The resultant solution was kept overnight in the flask sealed off completely for degassing.

4.2.1.2 S2 Solution
PSf (18 wt%) and PVP (1 wt%) were similarly dissolved in DMF as described earlier. The resultant solution was degassed overnight.

4.2.1.3 S3 Solution
PAN homopolymer (supplied by M/s. Technorbital Advanced Materials Pvt. Ltd., Kanpur, India) was dissolved in DMF at around 60°C in a beaker with the help of an external mechanical stirrer. The time taken for entire dissolution was 2 hours.

4.2.2 Membrane Synthesis
Even though lacquer synthesis was slightly different, flat-sheet membrane synthesis follows identical methodology for all lacquers. The methodology found a brief mention in Chapter 2; however, here we discuss it in more detail. The first step involves placing of a nonwoven polyester fabric support (118 ± 22.8 µm thickness supplied by M/s. Hollytex, India Inc., New York, USA) over a glass plate. The lacquer is poured over the edge of the fabric and a doctor's blade (fabricated and supplied by Gurpreet Engg. Works, Kanpur, India) is used to draw the solution over the fabric, casting a thin film of uniform thickness as set by the doctor's blade. For this study, it was set at 150 µm. Once cast, the film was immediately immersed in a distilled water gelation

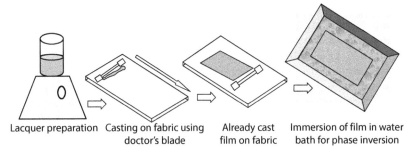

Lacquer preparation   Casting on fabric using   Already cast   Immersion of film in water
doctor's blade   film on fabric   bath for phase inversion

Figure 4.1   Membrane casting steps.

bath overnight to complete the phase inversion. S1 and S3 were cast, and the membranes obtained were kept for analysis. S2 needed further treatment. The cast membrane was heated at 75°C for 10 minutes. It was then immersed in 2% aqueous solution of MPD (supplied by M/s. Merck [India] Ltd.) for 5 minutes and then dried in air for 15 minutes. Then, it was immersed in 0.1% TMC (dissolved in hexane) solution for 5 minutes and air-dried again for 15 minutes. Finally, it was immersed in distilled water and left overnight. The steps are depicted in Figure 4.1.

4.2.3 Physical Characterization of Membranes
4.2.3.1 Permeability and MWCO
Permeability of a membrane indicates the water flux through it per unit area per unit time per unit of pressure applied. The higher the permeability, the more porous the membrane. MWCO indicates the particular molecular weight above which solutes will be retained completely by the membrane and below which value solutes will be allowed to permeate through the membrane. Both permeability and MWCO were calculated with the help of a stirred batch cell setup.[2] The first step involves "compacting" the membranes at a pressure higher than operating pressure ranges encountered by it during operations (discussed later). Hence, it was compacted at 690 kPa for 3 hours. Using distilled water, flux was noted at five different applied pressures. The permeate flux was calculated by

$$v_w = \frac{Q}{A\Delta T} \tag{4.1}$$

where
    v_w is pure water flux
    Q is the volumetric flow rate of permeating water
    A is the effective filtration area (33.16 cm^2)
    ΔT is the sampling time

A plot of v_w against applied transmembrane pressure results in a straight line passing through the origin. The slope of this line yielded the hydraulic permeability of the membrane.

In the same setup, the MWCO was also found out for the three membranes. A wide range of molecular weights of neutral solutes (1,000, 4,000, 10,000, 20,000, 100,000 Da), viz., PEG (supplied by M/s. S R Ltd., Mumbai, India), were taken. Dextran (molecular weight 70,000 Da, supplied by M/s. Sigma Aldrich, USA) was also considered for this purpose. Solutions of a concentration of 10 kg/m³ were prepared for each solute by dissolving it in distilled water and fed to the batch cell.[1] At high stirring speeds (2000 rpm) and

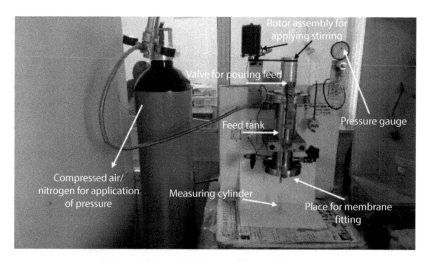

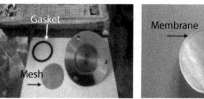

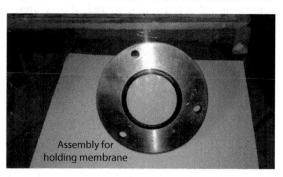

Figure 4.2   Batch cell and assembly for membrane placement.

low transmembrane pressure drop (70 kPa), the permeate from the membrane was collected in intervals of 5 minutes. The percentage rejection (%R) was measured:

$$R = \left(1 - \frac{C_P}{C_F}\right) \times 100\% \qquad (4.2)$$

where C_F and C_P are solute concentrations in the feed and permeate, respectively. Rejection of various solutes as calculated by Equation 4.2 was plotted against the molecular weight of solutes, and a sigmoidal curve was obtained. The molecular weight corresponding to the point of 90% rejection was the MWCO of the membrane.

A typical batch cell setup is described in Figure 4.2. It consists of a nitrogen/compressed air supply line. This air is used to apply pressure over the membrane surface for filtration to occur. The cell is fit with a stirrer. The feed tank has a pressure gauge for indicating the applied pressure, a valve for pouring in the feed, the feed tank, and a flanged fitting for membrane placement. A membrane of appropriate geometry and dimension is cut to fit into the flange. This is placed over a mesh and finally a gasket is placed on the membrane to prevent leakage of feed to the permeate side.

PROBLEM 4.1
Find the hydraulic permeability of a newly cast membrane according to the method described earlier.

Solution:

Step 1: Take the cast membrane and cut out a circular portion that will fit in the batch cell setup.

Step 2: Place the membrane and fill the batch setup with distilled water. Ensure that there is no leakage and apply a pressure 20% higher than what you would normally encounter for your operations with the membrane. This is called membrane compaction. The distilled water permeating through the membrane is collected in a beaker. This compaction is carried out for 120 minutes.

Step 3: Once compaction is over, the membrane is taken out, rinsed with water, and placed again in the batch cell. The setup is again filled with water. To measure the permeability, pressure is applied and water is collected in a measuring cylinder. The volume of water (mL) collected per unit time (seconds) is noted. A typical table (Table 4.1) can be generated. From the data, the flux can be calculated according to Equation 4.1.

The procedure was repeated for other pressures as well. A second table was generated for applied pressure and the obtained flux values.

The data in Table 4.2 are plotted and a straight-line fit is generated (Figure 4.3). The slope of the curve yields the hydraulic permeability of the membrane. In this example, it turns out to be 8.81×10^{-10} m/Pa s.

Table 4.1 Typical Permeability Data and Calculations							
Pressure Applied (Pa)	Volume of Water Collected (mL)	Time Taken (seconds)	ΔV (mL)	ΔT (seconds)	$\Delta V/\Delta T$	Flux (m³/m² s)	Average Flux
413,685	0	0	—	—	—	—	0.000312
	1	8	1	8	1.25×10^{-6}	0.000344	
	2	17	1	9	1.11×10^{-6}	0.000306	
	3	26	1	9	1.11×10	0.000306	
	4	35	1	9	1.11×10	0.000306	
	5	44	1	9	1.11×10	0.000306	
	6	53	1	9	1.11×10	0.000306	

Table 4.2 Applied Pressure and Obtained Flux	
Applied Pressure (Pa)	Obtained Flux (m³/m² s)
689,475	0.000551
551,580	0.00042
413,685	0.000312
275,790	0.000191

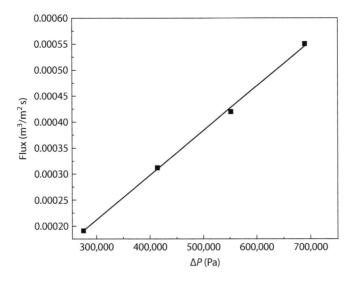

Figure 4.3 Plot of pressure versus flux.

PROBLEM 4.2
Find the molecular weight cutoff of the membrane.

Solution: As discussed previously, a variety of solutions were prepared from various neutral solutes dissolved in distilled water. The same setup was used (Figure 4.2). A 10 kg/m³ PEG 4000 solution was prepared as described previously. The refractive index (RI) of the PEG feed solution was recorded, and a calibration curve was prepared indicating concentration versus RI. The solution was poured into the feed tank and pressure (20 kPa) was applied. The permeate side concentration was measured by taking samples and using the calibrating curve. Now, if feed concentration (C_F) is 10,000 ppm and permeate concentration is (C_P) is 5,000 ppm, %R is 1–0.5 = 50%. A table (Table 4.3) is prepared for rejection against the various molecular weights.

The data were plotted as in Figure 4.4, and a sigmoidal function was fitted. The point that represents 90% rejection is the MWCO, which in this case is 40,000 Da or 40 kDa. So, any molecule above 40 kDa will be retained by the membrane and any molecule less than 40 kDa will permeate through.

Table 4.3 %R versus Molecular Weight	
Molecular Weight	%R
100,000	100
75,000	100
35,000	75
20,000	50

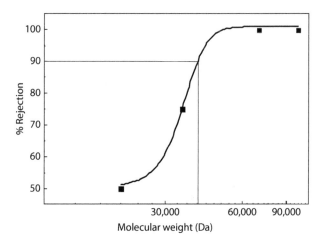

Figure 4.4   Molecular weight cutoff of the membrane.

4.2.3.2 Porosity of the Membranes

The percentage porosity was calculated by measuring the difference between the dry and wet weight of the cast membranes. Specific dimensions of membranes were cut (2 cm × 2 cm), immersed in distilled water, and taken out in 5 minutes. After drying off the superficial water, the wet weight was measured (w_0) and then placed in an air-circulating oven for 24 hours at 60°C and was subsequently placed in a vacuum oven for further drying. The dry weights (w_i) were then measured and the porosity was calculated[3]:

$$\varepsilon = \frac{w_0 - w_i}{\rho_w Al} \times 100\% \tag{4.3}$$

where
    ε is membrane porosity
    A is the area of the membrane
    l is membrane thickness
    ρ_w is water density

PROBLEM 4.3
Determine the porosity of the membrane in the previous example.

Solution: A membrane was cast and cut into small pieces, each of dimension 20 mm × 20 mm. The thickness measured by using a screw gauge was 0.3196 mm. It was then immersed in water, and the wet weight was recorded as 0.00017 kg. As described earlier, it was then dried and the dry weight was recorded as 0.000071 kg. The density of water is 1000 kg/m³. Hence, porosity (ε) is

$$\varepsilon = \frac{0.00017 - 0.000071}{0.00012784} \times 100 = 77\%$$

4.2.3.3 Tensile Strength

The mechanical strength of membranes was evaluated by using a universal testing machine (UTM), procured from M/s. Tinius Olsen Ltd., Redhill, England (model H50KS). A constant strain rate of 20 mm/min was maintained at 25°C. A typical UTM machine is depicted in Figure 4.5a. Basic components of the machine are a load cell, a moving crosshead, and grips to hold the sample (Figure 4.5b). The crosshead movement is controlled at a precise rate, thereby applying a tensile or compressive load to the specimen (Figure 4.5c). The load applied on the load cell is mostly hydraulic or gear-driven, although in some rudimentary machines, it can be applied by air piston and cylinders. Gear-driven machines are generally designed for crosshead speeds up to 0.001–500 mm/min. Screw-driven machines are one-, two-, or four-screw designs. In order to avoid twisting of the sample because of screw action, one screw has a left-hand thread while other has a right-hand thread. In order to provide lateral stability and alignment, screws are supported in bearings at each end. A range of crosshead speeds can be achieved by changing the gear ratio and

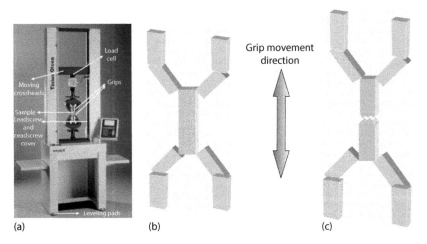

(a) (b) (c)

Figure 4.5 UTM testing of membranes (a) UTM machine, (b) sample under tensile stress, and (c) sample yielding with increasing tensile stress.

the speed of the electric motor. High-end sensors are used to determine and maintain the real crosshead speed within 0.125% of the user input speed.

The sample is placed in grips, and a constant tensile load is applied and elongation is recorded. Elongation (e) is measured according to[5]

$$e = \frac{L - L_0}{L_0} = \frac{\Delta L}{L_0} \tag{4.4}$$

and stress (σ) is measured according to[5]

$$\sigma = \frac{F}{A} \tag{4.5}$$

where
   L is the final length
   L_0 is the initial length
   ΔL is the change in length
   F is the force applied
   A is the area on which F is applied

Once the machine generates this data, it is plotted as a typical stress–strain curve as shown in Figure 4.6.

The linear portion of the curve is the region where the material follows Hooke's law, that is, stress (σ) applied is proportional to the strain (e) generated within the material, and the proportionality constant is Young's modulus (E) of the material. It is characteristic of a particular material.

$$\sigma = Ee \tag{4.6}$$

After further application of stress, beyond point (1) in Figure 4.6, the linearity between σ and e is lost and a point of ultimate strength is reached (2). Beyond

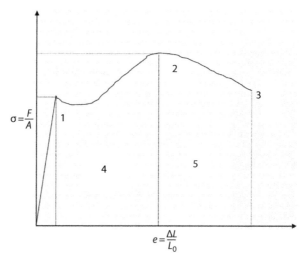

Figure 4.6 Stress–strain curve: (1) yield strength, (2) ultimate strength, (3) rupture, (4) strain hardening region, and (5) necking region.

this, necking occurs and the material ruptures (3). The strain hardening region (4) is the regime where dislocation generation and movements occur within the crystal structure of the material. Necking (5) is the region where a neck develops and the material rapidly reaches the rupture point,[5] indicating the ultimate stress. In this study, we record the ultimate stress of the developed membranes.

4.2.3.4 Contact Angle

A goniometer (New Jersey, USA, Rame'-Hart, Model No. 200-F4) was used to measure the contact angle of the cast membranes using the sessile drop method.[4] At six different locations, the contact angle was measured and the average value was reported.

In order to understand contact angle measurement, it is important to discuss surface tension. In a pure liquid, every molecule is pulled equally in all directions. The molecules near the surface, however, have unbalanced forces, since the surface does not have equal molecules in all directions. This is depicted in Figure 4.7a. Hence, the molecules maintain a configuration that will minimize its surface energy, and this is why droplets are round, since it yields minimum surface energy for a given volume.[6]

The way in which a liquid droplet interacts with a particular surface gives an idea about the wettability characteristic of the surface. This is measured by the contact angle of the droplet with the surface. This is depicted in Figure 4.7b. The contact angle is defined as the angle between the liquid–solid and liquid–vapor interfaces. From the figure, it is clear that a liquid drop on a surface can create three types of angles (θ). When $\theta > 90°$, it basically implies that the liquid droplet does not "like" the surface and would minimize its contact with it, forming a compact drop. Wetting in this case is unfavorable. The other extreme of such a case is when $\theta < 90°$, that is, when the drop maximizes its contact surface with the solid. In this case, it "likes" the surface and wetting is favorable. If the liquid is water, then $\theta > 90°$ implies "hydrophobic"

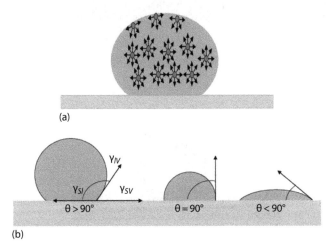

Figure 4.7 (a) Molecules interacting in a liquid droplet and (b) different surface wetting of droplets.

surfaces, and when $\theta > 150°$ the surface is "superhydrophobic." On the other hand, $\theta < 90°$ implies "hydrophilic" surfaces. In Greek, "gonia" means angle and "metron" means measure. A goniometer is used for measuring the liquid contact angle on a substrate. A typical goniometer is depicted in Figure 4.8a.

It consists of a horizontal platform where the sample is to be placed. Both the horizontal position (and focus) and the vertical height adjustment are maintained with the help of screws that can move the platform in front–back and up–down directions. A micrometer pipette is used to place a small droplet carefully on the surface of the substrate. The illumination device throws light over the sample, whereas the telescopic eyepiece with a protractor measures the interfacial contact angle, and this is recorded with the help of proprietary software. A typical droplet image of a goniometer is depicted in Figure 4.8b.

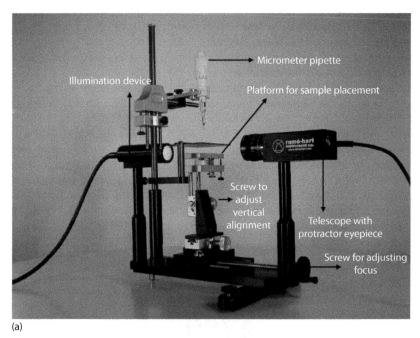

(a)

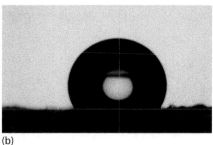

(b)

Figure 4.8 (a) Goniometer for measuring contact angle and (b) typical goniometer image of a droplet.

4.2.3.5 Membrane Morphology

The membranes were dried overnight in a desiccator before being dipped in liquid nitrogen and fractured. They were then gold-coated and placed on stubs. They were viewed at proper magnification under a scanning electron microscope (model: ESM-5800, JEOL, Japan).

A scanning electron microscope (SEM) is one of the most sophisticated instruments that is used for analyzing membrane structure and morphology. A simplified diagram of the operating principle of a SEM and a typical SEM machine are depicted in Figure 4.9.[7] It consists of an electron gun that emits an electron beam with energy levels varying from 0.2 to 40 keV. The beam is focused to a diameter of about 0.4–5 nm with the help of a magnetic lens. It then passes through a scanning coil that deflects the beam along the X- and Y-axes in order to carry out a rectangular pattern scanning of the sample. Finally, once the beam hits the sample, three types of emissions occur. Due to elastic collision, high-energy electrons are emitted from the sample. Inelastic collision leads to emission of secondary electrons, and finally emission of electromagnetic radiation occurs. All three types of radiation are captured by specialized detectors, and amplifiers are used to amplify weak signals. This results in a distribution map of scattering intensity that is obtained in the form of digital images. In order to focus the beams, a vacuum below 10^{-4} Torr is maintained. Most SEM machines operate at 10^{-6} Torr or higher. The magnification of SEM machines can vary within 6 orders of magnitude, ranging from 10 to 500,000 times. In order to visualize the samples, they have to be conducting and for this reason nonconducting polymeric samples like membranes need to be gold-coated. The SEM image of a typical dialysis hollow fiber and a flat-sheet membrane have been discussed in Chapter 2.[7]

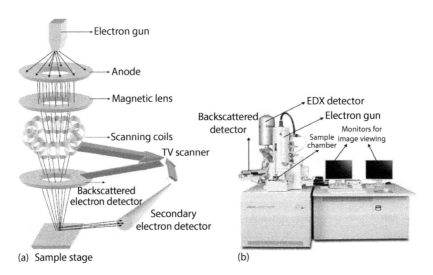

Figure 4.9   (a) SEM principle and (b) typical SEM machine.

4.2.3.6 Surface Charge Measurement

An electroultrafiltration cell was used to measure the surface charge of the membranes.[8] The various operating conditions maintained for the experiment were temperature = 298 ± 2.0 K, transmembrane pressure = 0–2 bar, solution pH = 7.4, NaCl concentration = 0.01 M, and cross-flow velocity of 0.12 m/s. The streaming potential was determined at the point where the net current is zero at an applied potential difference (ΔV) and pressure differential (ΔP). The streaming potential was determined from the slope of the plot between ΔV versus ΔP. The pertinent equations are

$$V_P = \left(\frac{\Delta V}{\Delta P} \right)_{I=0} \tag{4.7}$$

$$\zeta = \frac{V_P \mu \lambda}{\varepsilon_0 D_i} \tag{4.8}$$

where
  ζ is the membrane zeta potential
  ε_0 is the permittivity in vacuum
  D_i is the dielectric constant of the medium
  μ and λ are the viscosity and conductivity of the feed solution, respectively

In order to understand how membrane charge can be measured with the help of the theory discussed earlier, a few concepts need to be discussed beforehand. If a negatively charged particle is considered in a solution consisting of both positively and negatively charged particles, then an important phenomenon occurs, as described in Figure 4.10. Around the negatively charged particle, positive particles get attracted and form a layer that is called the Stern layer. Beyond the Stern layer, there is a diffused layer. The diffused layer consists of negative and positive ions that are in equilibrium with the Stern layer. In other words, the Stern layer exerts influence on the ions in the solution only until the diffused layer and the ions move if the particle moves under certain conditions like gravity or applied voltage. Actually, somewhere within the diffused layer lies the "slipping plane," where the influence of the particle ceases to exist and the potential between the particle and the slipping plane is called zeta potential (Figure 4.10a). However, considering the difficulty in determining the location of the slipping plane, the zeta potential is approximated as the potential of the Stern layer. If a membrane surface is charged, then it adsorbs oppositely charged particles on its surface, thereby altering the zeta potential. Measuring this gives an idea about the charge on the membrane surface. This is carried out with help of an electroultrafiltration unit as depicted in Figure 4.10b. It consists of a feed tank, from which the feed is pumped to the electroultrafiltration cell. There is a bypass, and a flow control valve helps in controlling the flow rate to the cell. The cell consists of a feed inlet and a feed outlet, which is the retentate stream (Figure 4.10c). The membrane is sandwiched between a stainless steel support and a platinum-coated titanium plate (Figure 4.10d). While the former forms the cathode, the platinum-coated titanium plate forms the anode.

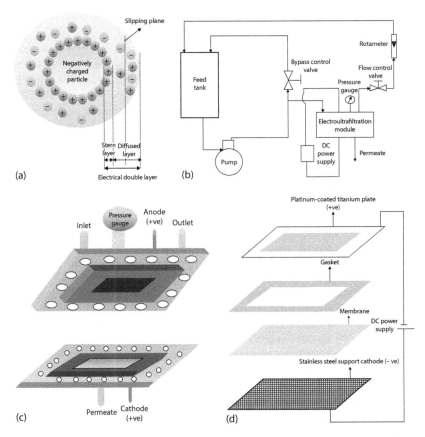

Figure 4.10   (a) Stern layer, slipping plane, and diffused layer; (b) electroultrafiltration unit; (c) construction of electroultrafiltration unit; and (d) components of
the unit for measuring charge.

An external electric field is applied from a regulated DC power supply across
the membrane. The retentate stream is recycled back to the feed tank, and the
permeate is collected from the bottom. The feed flows tangentially to the membrane surface through a thin channel 37 cm in length, 3.6 cm in width, and
6.5 mm in height.

4.2.4 Cytocompatibility Evaluation

To measure the cytocompatibility of membranes, NIH3T3 (mouse embryonic
fibroblast cell line) cells were procured from the National Centre for Cell Science
(NCCS), Pune, India. The media where these cells were grown contained
alpha-modified essential media (αMEM) (12561-056, Invitrogen Life Sciences,
India) with 1% antibiotics, antimyotic solution (penicillin 100 µg/mL, strepto-
mycin 10 µg/mL, and amphotericin-B 25 µg/mL; A002A Himedia, India), and

10% fetal bovine serum (Himedia, India). The operating conditions were 37°C, 95% humidity, and 5% CO_2 (Heracell150i, Thermo, USA).

The membranes were cut into identical dimensions and were sterilized and soaked in cell culture media overnight. On each sample, about 1×10^4 cells/cm^2 were seeded along with the control in a 12-well cell culture plate. The required volume of media was added to each well and was cultured for a specific time corresponding to the specific assay type. All the assays were performed in triplicate, and the mean of the obtained results was reported. Commercially available Fresenius fibers were chosen as control for the experiments.

4.2.4.1 Cell Metabolic Activity

The cell metabolic activity gives an idea regarding leaching of toxic chemicals or the occurrence of any adverse reaction near or on the surface of the substrate. It is measured by using an MTT (3-[4,5-dimethylthiazol-2-yl]-2,5 diphenyltetrazolium bromide) dye reduction assay on days 3 and 7.[9]

On the respective day of the assay, cell-seeded membranes were rinsed with phosphate buffered saline (PBS). They were further incubated with 200 µL of 5 mg/mL MTT solution (M5655, Sigma), in the dark, under standard cell culture conditions. Metabolically active cells reduce pale yellow MTT reagent to soluble purple-colored formazan crystals. These crystals were then solubilized in dimethyl sulfoxide, and absorbance was measured at 570 nm on a microplate reader (Recorders and Medicare Systems, India). The metabolic activity was correlated with the increase in absorbance values, since absorbance is proportional to the number of living and growing cells.

4.2.4.2 Cell Proliferation Assay

To evaluate cell proliferation response of the seeded cells, DNA quantification assay was carried out on days 3 and 7. The DNA content of the cells on the respective days was measured by DNA Quantitation Kit, Fluorescence Assay (DNAQF Sigma), based on standard protocols. The DNA of the cells binds to the fluorescent dye, bisBenzimide H 33258 (Hoechst 33258), of the assay. Then it was measured fluorometrically at an excitation wavelength of 350 nm and an emission wavelength of 460 nm. The standard DNA concentration curve was plotted with a standard solution of calf thymus DNA (D4810 Sigma).

4.2.4.3 Oxidative Stress Analysis

The dichloro-dihydrofuran-fluorescein diacetate (DCFH-DA) assay was utilized to measure the levels of oxidative stress generated in the cellular microenvironment during exposure to the membranes.[10] PBS was used to rinse the cell-seeded sample wells and incubated with 1 mM methanolic DCFH-DA solution (Sigma-Aldrich) at 37°C for 1 hour. Oxidative stress was estimated fluorometrically based on the cell interaction with DCFH-DA. The excitation wavelength was 785 nm and the emission wavelength was 530 nm as measured using a fluorescence spectrometer (Perkin Elmer, UK).

4.2.4.4 Estimation of Total Protein Content

The total protein content was estimated using two different methods.

4.2.4.4.1 Indirect Method

The bicinchoninic acid (BCA) protein assay[11] was used to estimate the total protein content in this method. NIH3T3 cells were seeded on samples on days 3 and 5. PBS-rinsed cell-seeded samples were incubated with BCA working solution (50 parts of BCA reagent with 1 part of 4% copper sulfate penta-hydrate, green-colored solution) at 37°C for 30 minutes. The mechanism involved in this is the reduction of free amino acids, which form a crimson-colored complex with BCA. The concentration of this was measured by measuring absorbance at 562 nm on a microplate reader (Recorders and Medicare Systems, India). The standard protein concentration curve was plotted with a known concentration of bovine serum albumin (BSA).

4.2.4.4.2 Direct Method

Three types of proteins, viz., BSA (5 g/dL), human γ-globulin (1.5 g/dL), and human fibrinogen (0.45 g/dL), were dissolved in phosphate buffered solution at 37°C for 2 hours. The samples were rinsed with PBS three times and were kept in 1 wt% aqueous solution of sodium dodecyl sulfate (SDS) for 60 minutes at room temperature on a shaker. The adsorbed proteins were removed from the samples and measured using a BCA protein assay.[11]

4.2.4.5 Cell Attachment and Morphology

SEM analyses of cells attached on samples were carried out on days 3 and 7.[12] The samples were rinsed gently with PBS, fixed by 4% paraformaldehyde at 37°C, and dehydrated with gradient ethanol solution and vacuum-dried overnight. The samples were then gold-coated for viewing under SEM.

For fluorescent imaging, cell-seeded samples were washed thrice with PBS and cells were fixed with 4% paraformaldehyde followed by permeabilization of cells using cell lysis solution (0.1% triton X in PBS). Cells fixed on the film were stained with the Hoechst dye (H1399, Invitrogen Life Sciences) according to the manufacturer's instructions. Images were acquired with Axio Observer Z1 (Carl Zeiss, Germany).

4.2.5 Hemocompatibility Evaluation

Whole blood was collected from healthy donors in a polyethylene disposable syringe containing 4.9% citrate–phosphate–dextrose–adenine (CPDA) solution. An anticoagulant was added to the blood, and the following tests were performed. Sample size was kept similar in each test and equilibrated with normal saline via incubation for 1 hour before each test. All the assays were performed in triplicate, and their mean was reported.

4.2.5.1 Hemolysis Assay

Red blood cell (RBC) compatibility was estimated with a hemolysis assay. Normal saline and 1% Triton X-100 solution were used as positive and negative controls. Saline-equilibrated samples were immersed in blood and incubated for 1 hour at 37°C, 95% humidity, and 5% CO_2 (Heracell150i, Thermo, USA). Any possible lysis of RBCs was quantified measuring optical density at 540 nm.

4.2.5.2 Blood Cell Aggregation

A blood cell aggregation study was carried out to measure the changes in the surface property of blood cells. Freshly collected blood was centrifuged at 700 rpm and the collected pellet was resuspended with normal saline in 1:9 volume ratios; 100 µL of this solution was mixed with 600 µL of normal saline. For RBC aggregation study, equal sizes of membranes were incubated with prepared suspension for 1 hour at 37°C. A white blood cell (WBC) aggregation study was carried out by isolating WBCs from uncoagulated freshly isolated blood by the Ficoll-Paque mononuclear cell isolation principle using HiSep™ LSM-1077 (Himedia) based on manufacturer's instructions. They were then mixed with normal saline and incubated with membranes as previously. After incubation, the cell suspension was smeared on a glass slide and observed under the microscope (Axio Observer Z1Carl Zeiss, Germany).

4.2.5.3 Platelet Adhesion

Platelet-rich plasma (PRP) was collected from fresh blood by centrifuging at 1500 rpm. Samples were pre-equilibrated in saline and incubated with PRP blood for 2 hours at 37°C, 95% humidity, and 5% CO_2 (Heracell150i, Thermo, USA). The samples were then rinsed with normal saline and fixed by 4% paraformaldehyde at 37°C. They were then further dehydrated with gradient ethanol solution and vacuum-dried overnight. Before optical characterization under SEM, the samples were gold-coated (polaron, UK).

4.2.5.4 Thrombus Formation

Degree of thrombosis (DOT) is an index to measure thrombus formation. Saline-equilibrated samples were incubated with freshly collected whole human blood in a 24-well plate for 2 hours at 37°C, 95% humidity, and 5% CO_2 (Heracell150i, Thermo, USA). Then the samples were rinsed gently three times with normal saline, fixed by 4% paraformaldehyde at 37°C. They were dehydrated with gradient ethanol solution and were vacuum-dried overnight. DOT was measured as[13]

$$DOT = \frac{W_t - W_d}{W_d} \qquad (4.9)$$

where
  W_t is the weight of the blood-treated sample
  W_d is the dry weight of the sample before blood treatment

4.2.6 Uremic Toxin Transport Capability

Even though a particular dialysis material may qualify as an excellent "biocompatible" material, yet its most important function is its capability to transport uremic toxins. The last test to which the three membranes were subjected to, in order to select the best material for dialysis, was investigation of the transport rates of urea and creatinine. A setup, as described in Figure 4.11, was used. The cast membranes were cut and fitted into a cross-flow membrane module set up as shown in the figure. A feed tank (a) contained the uremic toxins (urea and creatinine) dissolved in distilled water, which was driven by

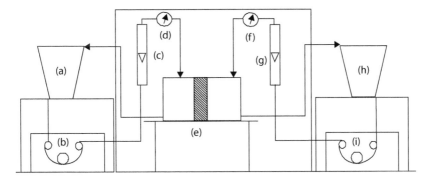

Figure 4.11 Experimental setup for measuring urea and creatinine permeances: a—feed tank; b—peristaltic pump; c—rotameter; d—pressure gauge (sphygmomanometer); e—cross-flow membrane module; f—pressure gauge (sphygmomanometer); g—rotameter; h—dialysate tank; i—peristaltic pump. (From Roy, A., et al., *RSC Adv.*, 5(10), 7023–7034, 2015. Reproduced by permission of The Royal Society of Chemistry.)

the peristaltic pump (b). The feed passed through the membrane module via the rotameter (c), which was used to measure the flow rate. A similar circuit was used in the dialysate side with distilled water as the dialysate. The flow rate of the feed side was maintained at 250 mL/min and that of the dialysate side at 250 and 500 mL/min. Concentration of urea and creatinine in the feed was 500 and 20 mg/L, respectively.

4.3 Selection of Best Material
4.3.1 Selection on the Basis of Physical Properties
4.3.1.1 Permeability and MWCO

Based on the methodology discussed earlier, which was adopted to carry out the selection on the basis of cytocompatibility, hemocompatibility, and transport properties, this section discusses the results thus obtained and the selection rationale. Figure 4.12 depicts the contact angle and permeability of the three cast membranes.

Hydraulic permeability is often attributed to the hydrophilicity of membranes. If the membrane is hydrophilic, it tends to maintain a thin film of water over its surface, thereby reducing the adsorption of solutes. Hydrophilicity is induced on a surface by increasing the number of hydrogen bond interaction sites with the addition of water.[15] Decrease in adsorption of solutes or proteins is very important for applications like dialysis, since adsorption of proteins during dialysis would lead to complications like decrease in transport efficiency of toxins and complement activation. It is observed that S3 has the least permeability of all the three membranes (0.2×10^{-10} m/s Pa), S1 has the highest permeability (1.4×10^{-10} m/s Pa), and the permeability of S2 lies between

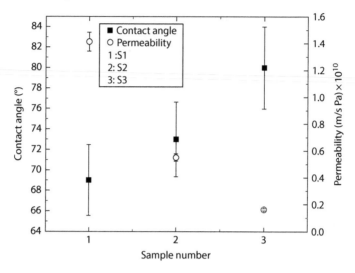

Figure 4.12 Permeability and contact angle of the three membranes. (From Roy et al., *RSC Adv.*, 5, 7023–7034, 2015. Reproduced by permission of The Royal Society of Chemistry.)

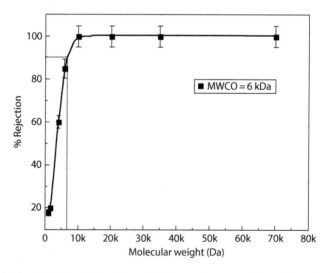

Figure 4.13 MWCO of S1, S2, and S3 membranes. (From Roy et al., *RSC Adv.*, 5, 7023–7034, 2015. Reproduced by permission of The Royal Society of Chemistry.)

the two (S2 is 0.6×10^{-10} m/s Pa). This can be attributed to the variation in the contact angle. S1 has the least contact angle of 69°, S3 is the most hydrophobic of the membranes (80°), and S2 is somewhere between the two with a contact angle of 73°. The MWCO of all the three membranes is 6 kDa as depicted in Figure 4.13.

The increase in hydrophilicity of the S1 membrane as compared to S2 and S3 is due to the addition of two hydrophilic polymers, viz., PVP and PEG. The separation characteristics of the membranes have been captured in terms of permeability and MWCO characteristics. It is clear that all three membranes have the same MWCO, which means molecules of the same size can permeate through the membrane. Hence, hydraulic permeabilities are expected to play the governing role in transport properties. However, other specific physical properties that cause variation of the membrane structures manifesting in variation of permeabilities is discussed as follows.

4.3.1.2 Porosity, Surface Morphology, Tensile Strength, and Surface Charge

The cross-sectional SEM images of the membranes are depicted in Figure 4.14a through c. It is seen that the basic structure of all three membranes is similar. They have a selective skin layer followed by a porous sublayer and a spongy bottom layer (Figure 4.14d through f). A close inspection would reveal that the skin layer is almost similar for all three membranes, and this explains the similarity in MWCO. However, they vary to some degree in their porous sublayer structure. It was discussed earlier that the S1 membrane is synthesized from a blend of PSf, PVP, and PEG. Revisiting the fundamentals of the phase inversion mechanism, it can be understood that a hydrophilic polymer in a membrane matrix will attract more water molecules toward it, leading to quicker demixing. This would induce more porosity in the membrane structure as is visible from Figure 4.14a through c where S1 has more macrovoids than S2 and S3. This is also expected to have an impact on other physical properties like porosity and tensile strength. This is shown in Figure 4.15. It is seen that S1 has the highest porosity and S3 the lowest. This understandably causes the tensile strength of the membranes to behave as a mirror reflection of the porosity.

Higher porosity results in lower tensile strength and vice versa. Hence, S1 has a porosity of 62%, S2 has an intermediate porosity of 60%, and S3 has a porosity of 54%, and their respective breaking stress values are 6, 7, and 11 MPa. The surface charges of S1 and S2 were nearly neutral (0.0–0.1 mV), while S3 was slightly negative in charge (−0.03 mV), which can be attributed to the presence of nitrile groups.

Thus, it is evident that the three membranes have similar MWCO, but S1 has higher hydrophilicity and hence is more fouling-resistant, which is desirable especially for dialysis applications. Higher permeability can be attributed to higher porosity of the membrane and although this decreases its mechanical strength, yet S1 can be selected amongst the three, since synthesizing a hollow-fiber dialysis membrane is the reason behind investigating the mechanical strength of the membranes.

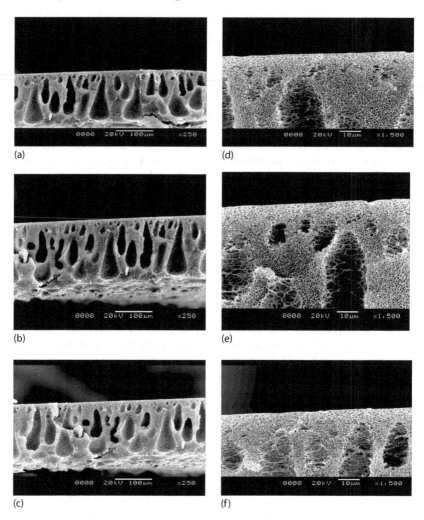

Figure 4.14   SEM images of the three membranes: (a) cross section of S1; (b) cross section of S2; (c) cross section of S3; (d) skin image of S1; (e) skin image of S2; and (f) skin image of S3. (From Roy et al., *RSC Adv.*, 5, 7023–7034, 2015. Reproduced by permission of The Royal Society of Chemistry.)

4.3.2 Selection on the Basis of Cytocompatibility
4.3.2.1 Metabolic Activity and Cell Proliferation

Figure 4.16 depicts the cell metabolic activity of the three membranes. The MTT assay indicated that S1 displayed higher cell growth (0.67 ± 0.05 and 1.06 ± 0.05 on days 3 and 7, respectively) on days 3 and 7 than either S2 or S3. S2 and S3, however, had almost similar growth kinetics (0.50 ± 0.03, 0.56 ± 0.03 on day 3 and 0.83 ± 0.04, 0.88 ± 0.04 on day 7, respectively). Interestingly, all three membranes had cell growth kinetics much higher than the control

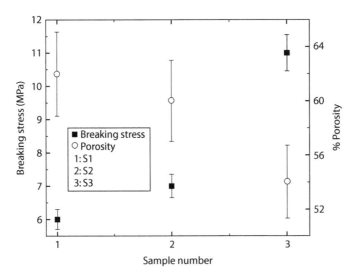

Figure 4.15 Tensile strength and porosity of the three membranes. (From Roy et al., *RSC Adv.*, 5, 7023–7034, 2015. Reproduced by permission of The Royal Society of Chemistry.)

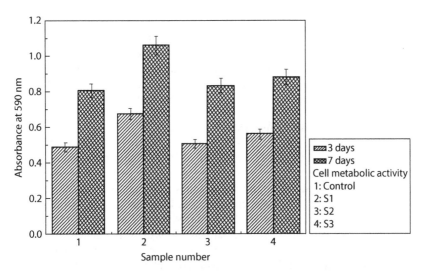

Figure 4.16 Cytocompatibility assays of the three membranes. (From Roy et al., *RSC Adv.*, 5, 7023–7034, 2015. Reproduced by permission of The Royal Society of Chemistry.)

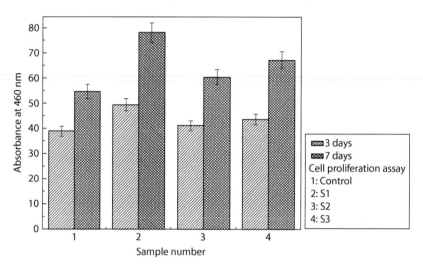

Figure 4.17 Cell proliferation assay of the three membranes. (From Roy et al., *RSC Adv.*, 5, 7023–7034, 2015. Reproduced by permission of The Royal Society of Chemistry.)

(0.48 ± 0.01 and 0.80 ± 0.01 on days 3 and 7). It can be concluded that all membranes qualified with regard to cytocompatibility and had favorable biological activity, with S1 being the most favorable of the three.

The DNA quantification assay results have been reported in Figure 4.17. It is evident again that the S1 membrane demonstrates higher cell proliferation rates than either S2 or S3. However, all three membranes have higher proliferation rates than the control. Absorbance value for S1-seeded cells was 49.4 ± 0.3 on day 3 of cell seeding followed by 78.08 ± 0.1 on day 7, whereas DNA-bound dye absorbance values for seeded cells on S2 and S3 were 41.16 ± 0.7, 43.59 ± 0.2 on day 3 and 60.46 ± 0.3, 67.15 ± 0.2 on day 7, respectively. All these findings are as expected since PSf–PVP is biocompatible in nature.[16]

4.3.2.2 Oxidative Stress Analysis

As described earlier, the DCFH-DA assay was used to quantify reactive oxygen species (ROS). DCFH-DA is oxidized to 20,70-dichlorofluorescein (DCF) by ROS formed near the membranes. Hence, the concentration of DCF would indicate ROS activity near the membranes. S2 and S3 initially displayed lower ROS activity but it was enhanced and similar to that of the control during incubation (Figure 4.18). S1 displayed much lower ROS activity with respect to control. This can be explained on basis of the general cytocompatibility of membranes. Interestingly, S2, a surface-modified membrane, exhibited comparable ROS activity with control and this can be attributed to phenylenediamine. Phenylenediamine derivatives have well-reported antioxidant activity[17] due to primarily three mechanisms: (1) free-radical scavenger ability,[18] (2) inhibition of oxidative glutamate toxicity, and (3) acting as

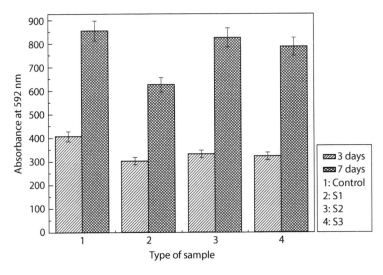

Figure 4.18 Oxidative stress analysis of membranes. (From Roy et al., *RSC Adv.*, 5, 7023–7034, 2015. Reproduced by permission of The Royal Society of Chemistry.)

peroxide decomposers by the eliminating oxidative catalyst to avoid further oxidation.[19]

4.3.2.3 Total Protein Content Estimation

Protein adsorption tests are vital for blood dialysis membranes. This is one component in the blood that can foul the membrane severely decreasing the dialysis efficiency, thus prolonging the dialysis time. As described, protein adsorption in this study was measured through direct (using human plasma proteins such as albumin, γ-globulin, and fibrinogen) and indirect methods (via NIH3T3 cell incubation).

It is well reported that protein adsorption is primarily due to hydrophobic surfaces. In an aqueous system, initial surface hydration of hydrophobic material governs the subsequent protein adsorption.[9] Hydrated protein molecules displace interfacial water via electrostatic interaction, thereby achieving thermodynamic equilibrium.[20,21] In the direct method, it was observed (Figure 4.19) that S3 exhibited the highest protein adsorption (30 μg/cm²) due to higher hydrophobicity of the membrane (contact angle 80°). Similarly, S2 and S1 membranes exhibit the protein adsorption results in accordance with their degrees of hydrophobicity. S2 (73°) has higher contact angle than S1 (69°) and hence exhibits higher protein adsorption (20 μg/cm²) than S1 (10 μg/cm²).

It is also to be noted that surface charges too play a significant role in protein adsorption; however, in this case these are minimal in magnitude and hence hydrophobicity dominates.[22,23] Moreover, the zwitterionic/mixed-charge

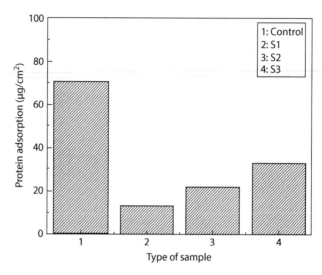

Figure 4.19 Protein adsorption of the membranes. (From Roy et al., *RSC Adv.*, 5, 7023–7034, 2015. Reproduced by permission of The Royal Society of Chemistry.)

hydration phenomenon does not exist here owing to high contact angle (73°) compared to zwitterionic surfaces (<20°). During the indirect protein adsorption study, similar results were obtained. S1 displayed the lowest protein adsorption profile compared to S2 and S3.

4.3.2.4 Cell Attachment and Morphology

The cell attachment and morphology of the membranes have been depicted in Figure 4.20. The membranes were viewed under both SEM and fluorescent microscope as depicted in Figure 4.20. The membranes exhibited less cell attachment initially (day 3), and S2 and S3 continued to show such behavior on day 7 as well. Cell attachment is a very complicated phenomenon and various factors (surface charge, surface roughness, hydrophilicity, etc.) play a role, and this is discussed in succeeding chapters. However, it is to be noted that all three membranes exhibited moderate cell adhesion characteristics. Again, it is observed in cytocompatibility assays that the S1 membrane performs better than S2 and S3. Thus, on the basis of physiological performance and cytocompatibility results, S1 is a better choice than S2 and S3.

4.3.3 Selection on the Basis of Hemocompatibility

4.3.3.1 Hemolysis Assay

The hemolysis assay results are depicted in Figure 4.21. For 2 hours, there was no significant change in hemolysis. All three membranes exhibited results comparable with the positive control, and there was no significant variation amongst them. Moreover, they exhibited less than 2% hemolysis suggesting

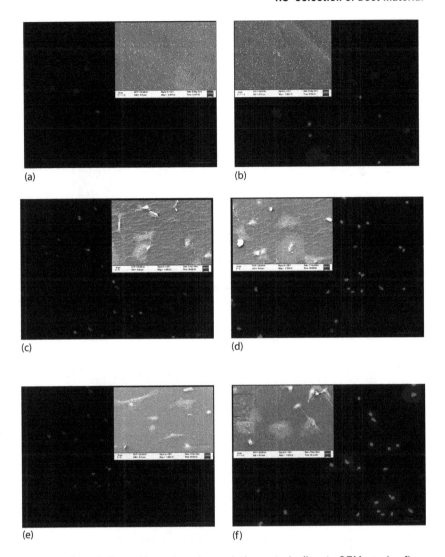

Figure 4.20   Cell attachment and morphology study (inset, SEM; main, fluorescent): (a) S1, 3 days; (b) S1, 7 days; (c) S2, 3 days; (d) S2, 7 days; (e) S3, 3 days; and (f) S3, 7 days. (From Roy et al., *RSC Adv.*, 5, 7023–7034, 2015. Reproduced by permission of The Royal Society of Chemistry.)

blood cell compatibility. Again, this behavior is further substantiated by the blood cell aggregation results in the next section.

4.3.3.2 Blood Cell Aggregation

As discussed previously, blood cell aggregation studies were carried out to investigate the hemocompatibility of synthesized membranes. There was no significant aggregation of RBCs observed (Figure 4.22). Few WBCs were observed

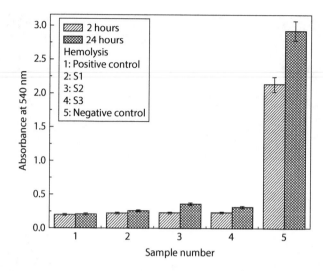

Figure 4.21 Hemolysis assay of the membranes. (From Roy et al., *RSC Adv.*, 5, 7023–7034, 2015. Reproduced by permission of The Royal Society of Chemistry.)

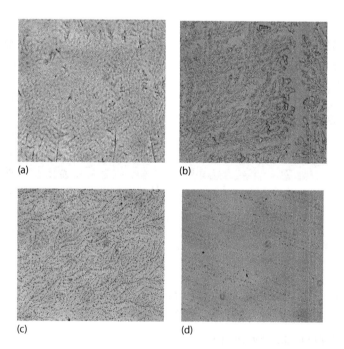

Figure 4.22 RBC aggregation on the three membranes: (a) control; (b) S1; (c) S2; and (d) S3. (From Roy et al., *RSC Adv.*, 5, 7023–7034, 2015. Reproduced by permission of The Royal Society of Chemistry.)

to rupture in S3, but they were not of much significance. These results further support hemolysis assay results. It has been reported that antifouling properties, surface charge, and smoothness control blood cell compatibility and aggregation activity during hemodialysis.[24] The S1 membrane shows promising hemocompatibility results due to incorporation of PVP-enhancing hydrophilicity, thereby reducing protein adsorption. Blending PVP with PEG in PSf reduces oxidative stress and controls the surface charge to near-neutral, which is also in agreement with reported literature.[25] Similarly, S2 membranes too exhibit comparable blood cell compatibility and due to the presence of phenylenediamine, ROS activity is reduced as has been discussed in previous sections. S3 has reduced hydrophilicity, and negative surface charge leads to cell aggregation.

4.3.3.3 Platelet Adhesion

Protein adsorption and platelet adhesion are closely interrelated phenomena. SEM images of the PRP-incubated membranes are depicted in Figure 4.23. It is observed that S3 experienced maximum platelet adhesion. S2 exhibited lower platelet adhesion than S3 with minimum morphological changes and S1 exhibited the least platelet adhesion amongst the three. The lower platelet adhesion is because of protein adsorption behavior. Ishihara et al. (1999) suggested that fibrinogen adsorption is a prerequisite for platelet adhesion.[26] The neutral surface charge leads to repulsion of proteins and negatively charged platelets.

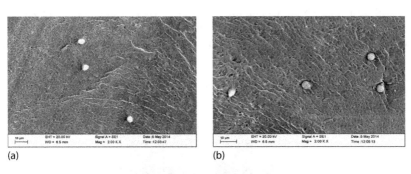

(a)   (b)

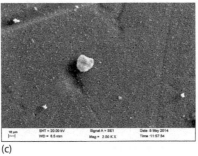

(c)

Figure 4.23   Platelet adhesion on the three membranes: (a) S1, (b) S2, and (c) S3. (From Roy et al., *RSC Adv.*, 5, 7023–7034, 2015. Reproduced by permission of The Royal Society of Chemistry.)

Tanaka et al. (2000) observed that not only fibrinogen adsorption but its conformational change also contributes toward platelet adhesion.[27] Fibrinogen, which is adsorbed on a surface but without any conformational change, never participates in platelet adhesion and activation without any conformational change. Thus, the S2 sample has lower fibrinogen adsorption without significant conformational changes compared to S3, resulting in low platelet adhesion and non-aggregation.

4.3.3.4 Thrombus Formation
The results obtained for thrombus formation are in line with platelet adhesion and protein adsorption (Figure 4.24). It is evident that S1 and S2 exhibit similar thrombus formation activity (0.31 and 0.37, respectively), and S3 exhibits highest DOT as 0.63. Thus, it is clear that on the basis of the hemocompatibility assay too S1 exhibits preferred performance over S2 and S3. The last criterion for selection of a hemodialysis membrane is the capability to transport uremic toxins, which is discussed in the next section.

4.3.4 Selection on the Basis of Uremic Toxin Transport Capability
The transport of uremic toxins is depicted in Figure 4.25. It is evident that the S1 membrane transports uremic toxins faster than either S2 or S3. This can be attributed to the permeability of the membranes, which of course is dependent on the porosity of the membranes, since each of their MWCO is equal. Hence, even if each of them can transport solutes of the same size, the rate of transport would depend on the porous structure and hence the permeability of the membranes. It is evident that S1 has higher permeability than S2 or S3; hence, it takes only 120 minutes to bring down the urea concentration

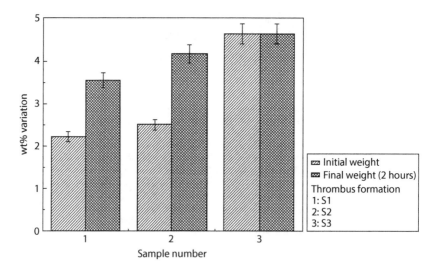

Figure 4.24 Thrombus formation on the three membranes. (From Roy et al., *RSC Adv.*, 5, 7023–7034, 2015. Reproduced by permission of The Royal Society of Chemistry.)

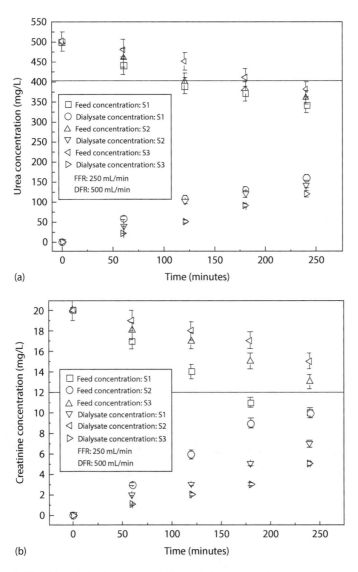

Figure 4.25   Uremic toxin transport through three membranes: (a) urea transport and (b) creatinine transport. *Note:* FFR, feed flow rate; DFR, dialysate flow rate. (From Roy et al., *RSC Adv.*, 5, 7023–7034, 2015. Reproduced by permission of The Royal Society of Chemistry.)

to 400 mg/L, whereas S2 takes slightly higher and it is significantly higher (180 minutes) in the case of S3. Similar is the case for creatinine as well.

4.4 Comparison of This Study with State-of-the-Art Literature

The comparison of this study with state-of-the-art literature has been summarized in Tables 4.4 and 4.5. The membrane characterizations have been mentioned in Table 4.4, and Table 4.5 illustrates the cytocompatibility and hemocompatibility comparisons. It is clear that the study in this chapter encompasses a total of nine parameters that have been used to evaluate the biocompatibility of the developed membranes.

4.5 Conclusion

From physical characterization, cytocompatibility, hemocompatibility, and uremic toxin transport capability performance, it is established that a blend membrane of Psf–PVP–PEG is suitable for dialysis applications. It not only results in the perfect MWCO and pore sizes along with permeability characteristics, but also more fundamental characteristics of the synthesized membranes are engineered in such a manner that it suits dialysis applications. Surface charge, hydrophilicity, and their mutual interplay, which determine the biological traits of cytocompatibility and hemocompatibility, have been studied. It is thus confirmed that a blend of the three polymers and their possible combinations should be used in the next and most challenging study and that is hollow-fiber membrane extrusions. This is discussed in the next chapter.

Table 4.4   Comparison of Membrane Physiological Properties with State-of-the-Art Literature

Membrane Properties	Present Work	State-of-the-Art Literature[13,16-21]
Hydraulic permeability (m/s Pa)	$0.2-1.4 \times 10^{-10}$	$3.12 \times 10^{-11} - 4.16 \times 10^{-10}$
Molecular weight cutoff (kDa)	6	—
Contact angle	68°–80°	56°–72°
Tensile strength (MPa)	6–11	—
Porosity (%)	53–62	—
Surface charge (mV)	−0.03 to 0.1	—
Surface morphology (SEM)	Studied	Studied

Table 4.5 Comparison of Membrane Cytocompatibility and Hemocompatibility with State-of-the-Art Literature

	Present Work	State-of-the-Art Literature					
		Li et al.[28]	Nie et al.[29]	Higuchi et al.[30]	Dahe et al.[31]	Shakaib et al.[32,33]	Lin et al.[13]
Material	PSf–PVP–PEG; surface-modified PSf; PAN homopolymer	PES blended with citric-acid-grafted polyurethane	Carbon-nanotube-grafted PES composite	Chemically modified PSf	PSf–vitamin E–TPGS composite	Polyamide and monosodium glutamate blend	PAN immobilized with chitosan and heparin
Cytocompatibility							
Metabolic activity (% better than control)	√√ (37%)	√√ (24%)	√√ (30%)	—	√√ (33%)	—	
Cell proliferation	√√	—	—	—	√√	—	—
Oxidative stress	√√	—	—	—	√√	—	—
Total protein adsorption (μg/cm²)	√√ 15–30	√√ 18-34	√√ 7-15	√√ 3-8	√√ 18-30	—	—

(Continued)

Table 4.5 (*Continued*) Comparison of Membrane Cytocompatibility and Hemocompatibility with State-of-the-Art Literature

	Present Work	State-of-the-Art Literature					
		Li et al.[28]	Nie et al.[29]	Higuchi et al.[30]	Dahe et al.[31]	Shakaib et al.[32,33]	Lin et al.[13]
Cell attachment and morphology (SEM and confocal imaging)	√√	—	—	—	√√	—	—
Hemocompatibility							
Hemolysis	√√	—	—	—	√√	—	—
Blood cell aggregation	√√	—	—	—	—	—	—
Platelet adhesion	√√	√√	√√	√√	√√	—	—
Thrombus formation	√√	—	—	—	√√	—	—

References

1. Roy, A. and De, S. 2014. Extraction of steviol glycosides using novel cellulose acetate pthalate (CAP)–Polyacrylonitrile blend membranes. *J. Food Eng.* 126: 7–16.
2. Rai, P., Majumdar, G.C., Dasgupta, S., and De, S. 2007. Modeling of permeate flux of synthetic fruit juice and mosambi juice (*Citrus sinensis* (L.) Osbeck) in stirred continuous ultrafiltration. *LWT—Food Sci. Technol.* 40: 1765–1773.
3. Li, J.F., Xu, Z.L., Yang, H., Yu, L.Y., and Liu, M. 2009. Effect of TiO_2 nanoparticles on the surface morphology and performance of microporous PES membrane. *Appl. Surf. Sci.* 255: 4725–4732.
4. Panda, S.R. and De, S. 2013. Role of polyethylene glycol with different solvents for tailor-made polysulfone membranes. *J. Polym. Res.* 20: 179–195.
5. Beer, F., Johnston, R., Dewolf, J., and Mazurek, D. 2009. *Mechanics of Materials.* New York: McGraw-Hill Companies.
6. Harvey, W.E. 1948. *Modern College Physics.* van Nostrand. ISBN 0-442-29401-8, New York.
7. Scanning Electron Microscope. n.d. Retrieved February 8, 2016, from https://www.purdue.edu/ehps/rem/rs/sem.htm.
8. Chatterjee, S. and De, S. 2014. Adsorptive removal of fluoride by activated alumina doped cellulose acetate phthalate (CAP) mixed matrix membrane. *Sep. Purif. Technol.* 125: 223–238.
9. Mosmann, T. 1983. Rapid colorimetric assay for cellular growth and survival: Application to proliferation and cytotoxicity assays. *J. Immunol. Methods* 65: 55–63.
10. Das, B., Dadhich, P., Pal, P., Srivas, K., Bankoti, K., and Dhara, S. 2014. Carbon nanodots from date molasses: New nanolights for the in vitro scavenging of reactive oxygen species. *J. Mater. Chem. B* 2(39): 6839–6847.
11. Smith, P.K., Krohn, R.I., Hermanson, G.T., Mallia, A.K., Gartner, F.H., Provenzano, M.D., Fujimoto, E.K., Goeke, N.M., Olson, B.J., and Klenk, D.C. 1985. Measurement of protein using bicinchoninic acid. *Anal. Biochem.* 85: 76–85.
12. Pati, F., Datta, P., Adhikari, B., Dhara, S., Ghosh, K., and Das Mohapatra, P.K. 2012. Collagen scaffolds derived from fresh water fish origin and their biocompatibility. *J. Biomed. Mater. Res. A* 100(4): 1068–1079.
13. W.-C., Lin, T.-Y., Liu, and M.-C. Yang. 2004. Hemocompatibility of polyacrylonitrile dialysis membrane immobilized with chitosan and heparin conjugate. *Biomaterials* 25: 1947–1957.
14. Roy, A., Dadhich, P., Dhara, S., and De, S. 2015. In vitro cytocompatibility and blood compatibility of polysulfone blend, surface-modified polysulfone and polyacrylonitrile membranes for hemodialysis. *RSC Adv.* 5(10): 7023–7034.
15. Sun, W., Liu, J., Chu, H., and Dong, B. 2013. Pretreatment and membrane hydrophilic modification to reduce membrane fouling. *Membranes* 3: 226–241.
16. Hayama, M., Yamamoto, K., Kohori, F., and Sakai, K. 2004. How polysulfone dialysis membranes containing polyvinylpyrrolidone achieve excellent biocompatibility? *J. Membr. Sci.* 234: 41–49.
17. Matsumoto, M., Yamaguchi, M., Yoshida, Y., Senuma, M., Takashima, H., Kawamura, T., and Hirose, A. 2013. An antioxidant, *N,N'*-diphenyl-*p*-phenylenediamine (DPPD), affects labor and delivery in rats: A 28-day repeated dose test and reproduction/developmental toxicity test. *Food. Chem. Toxicol.* 56: 290–296.

18. Chemicalland21. 2012. *N,N'*-Diphenyl-*p*-phenylenediamine. Accessed on 11/10/2014. http://www.chemicalland21.com/.
19. Satoh, T. and Izumi, M. 2007. Neuroprotective effects of phenylenediamine derivatives independent of an antioxidant pathway in neuronal HT22 cells. *Neurosci. Lett.* 418(1): 102–105.
20. Santore, M.M. and Wertz, C.F. 2005. Protein spreading kinetics at liquid–solid interfaces via an adsorption probe method. *Langmuir* 21: 10172–10178.
21. Nakanishi, K., Sakiyama, T., and Imamura, K. 2001. On the adsorption of proteins on solid surfaces, a common but very complicated phenomenon. *J. Biosci. Bioeng.* 91(3): 233–244.
22. Godjevargova, T. and Dimov, A. 1992. Permeability and protein adsorption of modified charged acrylonitrile copolymer membranes. *J. Membr. Sci.* 67: 283–287.
23. Wong, S.Y., Han, L., Timachova, K., Veselinovic, J., and Hyder, N. 2012. Drastically lowered protein adsorption on microbicidal hydrophobic/hydrophilic polyelectrolyte multilayers. *Biomacromolecules* 13(3): 719–726.
24. Wang, Z.-G., Wan, L.-S., and Xu, Z.-K. 2007. Surface engineering of polyacrylonitrile-based asymmetric membranes towards biomedical applications: An overview. *J. Membr. Sci.* 304: 8–23.
25. Hasler, C.R., Owen, G.R., Brunner, W., and Reinhart, W.H. 1998. Echinocytosis induced by haemodialysis. *Nephrol. Dial. Transplant.* 13: 3132–3137.
26. Ishihara, K., Fukumoto, K., Iwasaki, Y., and Nakabayashi, N. 1999. Modification of polysulfone with phospholipid polymer for improvement of the blood compatibility. Part 2. Protein adsorption and platelet adhesion. *Biomaterials* 20: 1553–1559.
27. Tanaka, M., Motomura, T., Kawada, M., Anzai, T., Kasori, Y., Shiroya, T., Shimura, K., Onishi, M., and Mochizuki, A. 2000. Blood compatible aspects of poly(2-methoxyethylacrylate)(PMEA)—Relationship between protein adsorption and platelet adhesion on PMEA surface. *Biomaterials* 21(14): 1471–1481.
28. Li, L., Cheng, C., Xiang, T., Tang, M., Zhao, W., Sun, S., and Zhao, C. 2012. Modification of polyethersulfone hemodialysis membrane by blending citric acid grafted polyurethane and its anticoagulant activity. *J. Membr. Sci.* 405: 261–274.
29. Nie, C., Ma, L., Xia, Y., He, C., Deng, J., Wang, L., Cheng, C., Sun, S., and Zhao, C. 2015. Novel heparin-mimicking polymer brush grafted carbon nanotube/PES composite membranes for safe and efficient blood purification. *J. Membr. Sci.* 475: 455–468.
30. Higuchi, A., Shirano, K., Harashima, M., Yoon, B.O., Hara, M., Hattori, M., and Imamura, K. 2002. Chemically modified polysulfone hollow fibers with vinylpyrrolidone having improved blood compatibility. *Biomaterials* 23(13): 2659–2666.
31. Dahe, G.J., Teotia, R.S., Kadam, S.S., and Bellare, J.R. 2011. The biocompatibility and separation performance of antioxidative polysulfone/vitamin E TPGS composite hollow fiber membranes. *Biomaterials* 32(2): 352–365.
32. Shakaib, M., Ahmed, I., Yunus, R.M., and Idris, A. 2013. Morphology and thermal characteristics of polyamide/monosodium glutamate membranes. *Int. J. Polym. Mater. Polym. Biomater.* 62(6): 345–350.
33. Shakaib, M., Ahmad, I., Yunus, R.M., and Noor, M.Z. 2014. Preparation and characterization of modified nylon 66 membrane for blood purification. *Int. J. Polym. Mater. Polym. Biomater.* 63(2): 80–85.

5

Spinning of Dialysis Grade Membranes

It has become appallingly obvious that our technology has exceeded our humanity.

—Albert Einstein

5.1 Introduction

The previous chapters laid down the foundation for the choice of polymers for synthesis of a suitable dialysis membrane. However, most of the literature seldom addresses the precise, focused question of "how to spin a hollow fiber using the polymers?" The engineering of a device capable of spinning hollow-fiber membranes is a challenge in itself. Few companies hold proprietary technology to spin hemodialysis grade hollow fibers. The process requirement (as discussed in Chapter 2) specifies that the dialysis hollow fibers be of 180–220 μm inner diameter with a thickness of 35–40 μm; 7,000–10,000 of such fibers are packed in a cartridge with a surface area of 1–1.5 m², maintaining a low residual volume of blood.

The extrusion of hemodialysis grade fibers poses a great challenge due to the delicate balance between the polymer flow rate, the water flow rate, the rheology of the polymer solution, along with the flow pattern of the fluids in the thin flow conduit. In this chapter, an indigenous technology that utilizes the polymeric blend solution discussed in the previous chapter and spin clinical grade hollow-fiber membranes is analyzed. The novelty of the design lies in using a low-cost, laboratory-fabricated syringe-in-syringe (SIS) assembly coupled with a complete process design.

This chapter is organized in the following manner: first, the role and interactions of the polymers in solution (as identified in the previous chapter) based on Fourier transform infrared (FTIR) analysis and rheological responses are analyzed. Second, the SIS assembly and process design for extrusion of hollow fibers are discussed. Finally, computational fluid dynamics (CFD) simulation for the flow patterns in the SIS assembly undertaken to visualize the flow patterns evolving in them and whether the new design has any effect of the flow at the outlet of the assembly is investigated.

5.2 Materials and Methods

5.2.1 Solution Preparation

The following polymer blend solutions (in DMF [dimethyl formamide] solvent) were synthesized using the technique mentioned in Chapter 4. These solutions were used for both FTIR and rheological analyses:

   a. PSf (18 wt%) and PEG (3 wt%)—referred to as PVP 0

   b. PSf (18 wt%), PEG (3 wt%), and PVP (3 wt%)—referred to as PVP 3

   c. PSf (18 wt%) and PVP (2 wt%)—referred to as PEG 0

   d. PSf (18 wt%), PVP (2 wt%), and PEG (3 wt%)—referred to as PEG 3

For hollow-fiber membrane spinning, the following compositions were used:

   a. PSf (18 wt%), PEG (3 wt%), and PVP (0 wt%)—referred to as 6 kDa

   b. PSf (18 wt%), PEG (3 wt%), and PVP (2 wt%)—referred to as 12 kDa

   c. PSf (18 wt%), PEG (3 wt%), and PVP (3 wt%)—referred to as 16 kDa

Flat-sheet membranes with the same composition as hollow fibers for contact angle analysis were prepared. The process is the same as discussed in Chapter 4.

5.2.2 FTIR Analysis

FTIR analysis was carried out to study the interpolymeric interactions in the blend solution. The solutions prepared were subjected to FTIR analysis in the attenuated total reflectance mode (ATR) mode (supplied by M/s. Perkin Elmer, CT; model: Spectrum 100). The analysis was conducted by studying the liquid film as a function of time by letting the solvent evaporate. The peaks obtained at different wavelengths indicated the functional groups present in subsequently formed membranes; 32 scans were carried out at a nominal resolution of $4 \, \text{cm}^{-1}$. Analysis of both the solution and the solid film (membrane) after complete evaporation of the solvent was undertaken.

In this regard, it is important to introduce the readers to the concept of FTIR analysis and its working principle. In Chapter 3, the working principle of a spectrophotometer was discussed. The working principle behind FTIR spectroscopy is slightly different. As shown in Figure 5.1a, a source generates a light, which, instead of passing through a monochromator (as in UV–VIS spectroscopy), passes through an interferometer. The light first passes through a beam splitter that sends the beam in two directions at right angles. One part goes through the stationary mirror and back to the beam splitter. The other part goes to a moving mirror, and it makes the total path length variable. The two

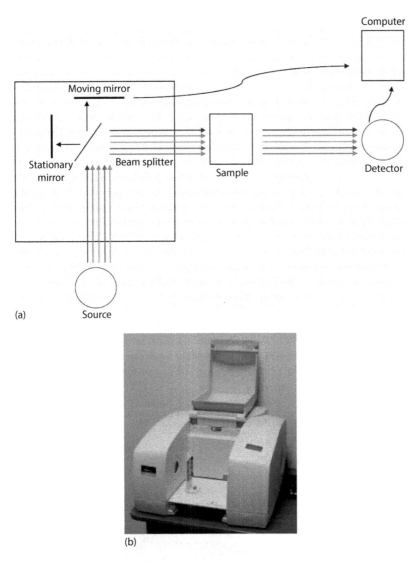

Figure 5.1 (a) Working principle of FTIR and (b) typical FTIR instrument.

beams recombine at the beam splitter and create constructive and destructive interference patterns. This recombined beam passes through the sample, which absorbs all the wavelengths and subtracts specific wavelengths from the inter-ference patterns. A laser beam is used for reference for the detector, and the wavelength from the sample is analyzed. The resulting energy is measured as a function of time, which is converted to frequency using the Fourier transform. Thus, an intensity versus frequency spectrum is generated. Each intensity peak at specific frequencies yields a designated functional group. Thus, functional groups present in the sample can be deciphered using FTIR analysis; moreover,

91

specific interactions between functional groups and variation in their energy levels can also be captured. Figure 5.1b depicts a typical FTIR instrument.

5.2.3 Rheology Study

The rheological behavior of the polymer solutions was measured to find the physical manifestations of the interactions occurring in the solution state. For this purpose, a rheometer supplied by M/s. Anton Paar (model: MCR 102) with cone and plate geometry, with a cone angle of 4° and a diameter of 40 mm, was used. The temperature was maintained at 25°C, and shear rates were varied from 0.01 to 1000 s^{-1}. To measure storage and loss modulus (G' and G'', respectively), shear stress of 5 Pa was selected, since the moduli exhibited shear stress–independent behavior between 1 and 10 Pa. In controlled strain tests, oscillatory shear rates were varied from 0.1 to 100 Hz. All rheological measurements were carried out with an accuracy of 25%.

5.2.3.1 Rheology Basics

In order to understand the succeeding sections, it is important to comprehend the basic concepts of rheology. The relevant topics to understand are

a. *Shear stress, shear rate, and Newton's law*:

To understand the basic principle of viscous forces, a simple model can be taken under consideration. As illustrated in Figure 5.2, suppose a thin film of liquid is confined between a flat surface and a plate and the plate is moved along the X-axis with a tangential force (F) acting over an area A. Then, shear stress (τ) is defined as the force acting over an area A:

$$\tau = \frac{F}{A} \tag{5.1}$$

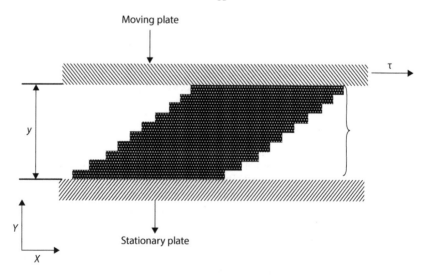

Figure 5.2 Flow between a stationary plate and a moving plate.

The layers of fluid, under such conditions, would experience a shear flow, with a relative velocity existing between them. It would behave like a deck of cards; hence, the layers in contact with the moving plate will move with maximum velocity and the layers farther away would try to remain stationary. A velocity gradient would exist along the Y-axis, and this is known as shear rate ($\dot{\gamma}$):

$$\dot{\gamma} = \frac{dv}{dy} \tag{5.2}$$

This is due to the resistance offered by the fluid against deformation or "flow" and is known as viscosity. Hence, Newton's law of viscosity can be stated as

$$\tau = \eta\dot{\gamma} \tag{5.3}$$

where η is the viscosity of the fluid.

b. *Dynamic viscosity and kinematic viscosity*:

There are two types of viscosities encountered while dealing with practical problems. The first is dynamic viscosity, which is obtained straight from the definition of Equation 5.3:

$$\eta = \frac{\tau}{\dot{\gamma}} \tag{5.4}$$

It is important to note the unit of dynamic viscosity, which is Pa s, and 1 Pa s = 1000 m Pa s; 1 m Pa s is also known as 1 centiPoise or cP. The viscosity of water is 1 cP.

Kinematic viscosity (v) is obtained when fluids are tested using capillary flowmeters, where flow is facilitated through gravity. It is obtained as

$$v = \frac{\eta}{\rho} \tag{5.5}$$

The various parameters on which η depend are (1) the physical or chemical nature of the substance, (2) temperature, (3) pressure, (4) $\dot{\gamma}$, and (5) time and the electric field.

c. *Types of fluid and their flow curves*:

Flow curves of various types of fluids are illustrated in Figure 5.3. Newtonian fluids are defined as per curve 1. It is seen that it is a straight line between shear stress and the shear rate, which indicates that the slope, that is, stress/rate, is a constant (Equation 5.4). This is the dynamic viscosity of the fluid. It can also be deduced from Figure 5.3 that increase in the shear rate does not increase the viscosity of the fluid. Any other fluid that has a viscosity change with the applied shear rate is classified as "non-Newtonian fluid." Curve 2 represents a dilatant fluid, which experiences an increase in

93

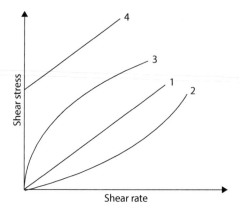

Figure 5.3 Types of flow curves: 1—Newtonian fluid; 2—dilatant fluid; 3—pseudo-plastic fluid; and 4—Bingham fluid.

viscosity (slope of curve increases) with the application of a higher shear rate and, hence, is referred to as shear thickening fluid. A typical example is plasticizers. Quite the opposite behavior is exhibited by pseudo-plastic fluids whose viscosity decreases with the shear rate and are hence called shear thinning fluids, with polymer solutions (as discussed in this book) being the classical example (curve 3). The behavior of all these fluids, including Newtonian, can be summarized with the power law for fluids:

$$\tau = k(\dot{\gamma})^n \tag{5.6}$$

where
k is the flow consistency index
n is the flow behavior index

It can be further simplified as

$$\tau = k(\dot{\gamma})^n = k\dot{\gamma}(\dot{\gamma})^{n-1} = \mu_{eff}\dot{\gamma} \tag{5.7}$$

where $k(\dot{\gamma})^{n-1} = \mu_{eff}$ is the effective viscosity. If $n < 1$ the fluid is pseudo-plastic, if $n > 1$ the fluid is dilatant, and if $n = 1$ the fluid is Newtonian.

The only other fluid not discussed thus far is curve 4, which is Bingham plastic fluid. The equation for curve 4 is

$$\tau = \tau_0 + k(\dot{\gamma})^n \tag{5.8}$$

where τ_0 is the yield stress. It signifies that the fluid does not begin to flow until some external agency has applied a minimum stress of τ_0, after which it follows Newtonian fluid behavior. A typical example is toothpaste.

A common characteristic of the fluids that have been discussed is that behavior is independent of time. However, there is another class of fluids whose behavior is dependent on time. They are thixotropic and rheopectic fluids. Their behavior is similar to that of shear thinning and shear thickening fluids, except that their viscosities change with time. Thixotropic fluids have decreasing viscosity over time, whereas rheopectic fluids have increasing viscosity with time. An example of a thixotropic fluid is clay and of a rheopectic fluid is gypsum pastes. A special type of fluid behavior that has not been described earlier is viscoelasticity. This is discussed as follows.

5.2.3.1.1 Polymer Solution and Viscoelasticity

With the basic background explained in the previous section, it is important to understand what membrane engineers deal with time to time. As has been discussed in previous chapters, membranes are synthesized from polymer solutions and the basic methodology to prepare the same has also been discussed. It is important to understand the basic fluid behavior of polymer solutions. Polymer solutions fall under the category of viscoelastic fluids. This has been the subject of study for long amongst polymer engineers, scientists, and physicists, and it continues to intrigue. Understanding viscoelasticity is challenging and an interesting example may illustrate why. The glass windows of the Cathedral of Chartres in France were manufactured 600 years ago. Initially, they were of uniform thickness. But after 600 years, the glass has "flown," making the top portion paper-thin and the bottom portion almost double its original thickness. This throws light on the fact that substances like glass, which apparently has solid characteristics, can behave like fluids under different conditions of shear rates, stress, or time. In order to understand what went wrong with the glass of the Cathedral of Chartres, it is imperative to discuss "springs and dashpots."

5.2.3.1.1.1 Viscoelasticity: What Is It and How to Model It?

Viscoelasticity can best be physically understood as the phenomenon that lends "memory" to viscosity! Viscous fluids have no memory, that is, once the applied stress or shear rate is removed the fluid returns to its original state. On the other hand, a perfect elastic solid, if stretched within elastic limits, has a memory of the load by retaining its deformed state even after the load is removed. These two extreme characteristics are understood by the dimensionless parameter called Deborah number (De), which is defined as

$$De = \frac{\text{Relaxation time of fluid } (\lambda)}{\text{Time scale of process}} \quad (5.9)$$

PROBLEM 5.1

A flow of fluid is occurring through a bed of particles. In one case, particle size is 50 mm and the velocity of the flow is 500 mm/s. In the other case, the particles size is 500 μm at the same velocity. Under what conditions will viscoelastic effects prevail if the relaxation time is 10 ms?

Solution: In the first case, the timescale of the process is $50/500 = 0.1$ s. Hence, the Deborah number is 0.1. In the second case, the timescale of the process is $500 \times 10^{-6}/500 \times 10^{-3} = 10^{-3}$ s. The Deborah number in the second case is 10. For an ideal elastic solid, the relaxation time of the fluid $\gg$ timescale, hence $De \to \infty$, whereas for a viscous fluid, the relaxation time is almost instantaneous, hence $De \to 0$.

Thus, from this example, it is clear that in the second case the viscoelastic effects will be more appreciable.

Thus, viscoelasticity can be defined as viscous and elastic effects prevailing simultaneously in a fluid. This behavior can be attributed to the structure of polymer chains and in order to model viscoelasticity, three major models have been in continuous use till date. They are the Maxwell, Voigt–Kelvin, and three-component mechanical models (Figure 5.4). The models consist of a spring and a dashpot system connected to each other either in series or in parallel. The spring imitates the elastic behavior and the dashpot imitates

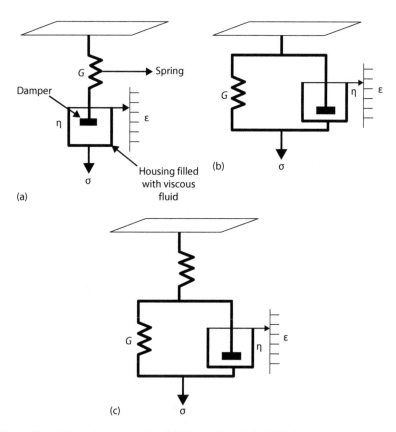

Figure 5.4 Viscoelastic models: (a) Maxwell model, (b) Voigt–Kelvin model, and (c) three-component mechanical model.

the viscous behavior. The dashpot consists of a damper housed in a viscous fluid. Together, they can be used to understand the relation between stress and strain and their responses in various viscoelastic systems.

5.2.3.1.1.2 Viscoelasticity: How to Measure It

Samples prepared were discussed in the previous sections. The measurements were carried out in an Anton Paar rheometer (MR 102 model) with cone and plate geometry, with a cone angle of 4° and a diameter of 40 mm. All the measurements were carried out at 25°C.

Standard rheometers that are used to measure viscosity and its variations due to applied shear stress, shear rates, temperature, and so on are available. These are also known as rotational viscometers. There are two types of rheometers: constant shear and constant rate. A typical rheometer machine (Figure 5.5a) has an air bearing–supported DC motor whose rotor is equipped with permanent magnets, whereas in the stator coils have opposite polarity, producing magnetic poles. The magnets in the rotor and the stator interact with each other and produce a flux of current that provides frictionless synchronous movement of the rotor. The torque produced by the motor is varied by varying the input current to the stator coil and, hence, can be measured. This torque is then used to measure shear rates, and the shear stress is measured using displacement of the upper measuring plate; thus, a curve between shear stress and shear rate yields viscosity of the sample. There are two types of arrangements to measure viscosity. These are the parallel plate and cone and plate (Figure 5.5a). Each has its own set of functionalities; the cone and plate are used for applying constant shear rate while the plate type is used for temperature-dependent tests. A typical rheometer is depicted in Figure 5.5b.

5.2.3.1.1.3 Some Relevant Definitions

In order to grasp the discussions in the following sections, it is important to introduce a few terms that are used in viscoelastic investigations. In order to understand the response of a non-Newtonian fluid, let us consider a strain that varies sinusoidally is applied, that is,

$$\gamma = \gamma_m \sin(\omega t) \tag{5.10}$$

Hence, the resultant shear stress would be

$$\tau = \eta \dot{\gamma} = \eta \gamma_m \omega \sin\left(\omega t + \frac{\pi}{2}\right) \tag{5.11}$$

Thus, it is clearly seen that the strain would lag behind the applied shear stress by a phase difference of π/2. The phase angle is denoted by δ and it varies from 0 to π/2; δ = 0 denotes purely viscous fluid and δ = π/2 indicates purely elastic behavior. Now, by Hookean elastic solid theory (as discussed in the previous chapter),

$$\tau = G\gamma \tag{5.12}$$

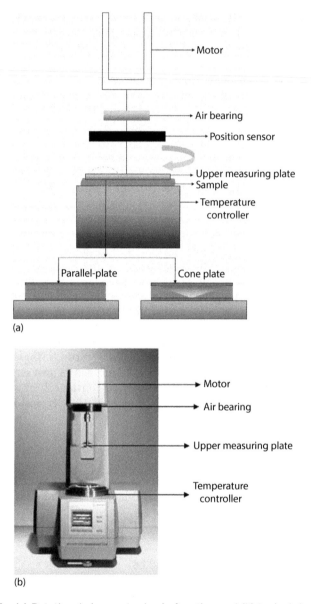

Figure 5.5   (a) Rotational viscometer basic function and (b) typical rheometer.

where G is the Young's modulus. According to linear viscoelastic theory, for any shear history, the applied shear stress to strain is a function of time alone and is independent of stress magnitude. Hence, the total strain experienced by a material is integral to the applied stresses throughout the history. Hence,

$$\tau(t) = \int_{-\infty}^{t} G(t-t')\dot{\gamma}(t')dt' \tag{5.13}$$

The shear stress and resultant strain can be expressed as complex quantities:

$$\gamma^* = \gamma_m e^{iwt} \tag{5.14}$$

$$\tau^* = \tau_m e^{i(wt+\delta)} \tag{5.15}$$

The complex modulus is thus defined as

$$G^* = \frac{\tau^*}{\gamma^*} \tag{5.16}$$

and

$$G^* = G' + iG'' \tag{5.17}$$

where G' and G'' are defined as

$$G' = \frac{\tau_m}{\gamma_m}\cos\delta \tag{5.18}$$

$$G'' = \frac{\tau_m}{\gamma_m}\sin\delta \tag{5.19}$$

G' is the dynamic rigidity. It is a measure of the energy stored and recovered per cycle of deformation. G'' is the loss modulus, which gives the measure of energy dissipated per cycle. In other words, it gives the extent of viscous behavior, and G' gives the extent of elastic behavior. Hence,

$$\delta = \tan^{-1}\frac{G''}{G'} \tag{5.20}$$

Similarly, dynamic viscosity can also be expressed as the summation of elastic and viscous parts:

$$\eta^* = \eta' + i\eta'' \tag{5.21}$$

With this basic knowledge, it will be interesting to understand the behavior of the polymer blend subject to rheological tests.

5.2.4 Membrane Characterization
5.2.4.1 Experimental Setup

The spun membranes were characterized in terms of their hydraulic permeability and molecular weight cutoff (MWCO), and their performance was analyzed using an experimental setup, the schematic of which is depicted in Figure 5.6. It consists of a laboratory-fabricated dialyzer using 300 spun dialysis fibers packed into a high-density polyethylene pipe and sealed at the ends with polyurethane resin. The feed is driven by a peristaltic pump from the feed tank. The feed flows through the dialyzer and is recycled back to the tank. The dialysate circuit consists of another peristaltic pump circulating

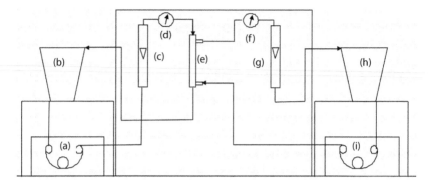

Figure 5.6 Experimental setup for measuring urea and creatinine permeances: a—peristaltic pump; b—feed tank; c—rotameter; d—pressure gauge (sphyg-momanometer); e—dialyzer cartridge; f—pressure gauge (sphygmomanometer); g—rotameter; h—dialysate tank; and i—peristaltic pump.

the dialysate (distilled water) through a rotameter and pressure gauge and is recycled back to the tank.

5.2.4.2 MWCO and Surface Morphology

The MWCO of the membranes was measured using the setup, with the dialysate circuit closed. The feed tank was filled with 10 kg/m³ of various polymeric solutions and a pressure of 10 mmHg was applied while maintaining a flow rate of 10 mL/min. The permeate coming out was collected, and the concentration was measured using a refractometer. The scanning electron microscope (SEM) images of the dialysis fibers were taken to study the surface morphology.

5.2.4.3 Permeability and Contact Angle

The permeability of the spun membranes was measured at various transmembrane pressures (TMPs) (200, 40, 60, and 80 mmHg). The contact angle of the flat-sheet membranes was determined using the sessile drop method as described in Chapter 4.

5.2.4.4 Porosity, Pore Side Distribution, and Tensile Strength

The porosity and tensile strength were measured as discussed previously (Chapter 4). The pore size distribution was calculated as reported in the literature.[1] The pore size distribution function is given as

$$\mathrm{PDD}(R) = \frac{1}{\sigma R \sqrt{2\pi}} \exp\left[-\frac{(\ln R - m)^2}{2\sigma^2}\right] \qquad (5.22)$$

where
    R is the pore diameter in nm
    σ and m are standard deviation and mean of distribution, respectively

From the MWCO curves, the parameter U_R can be calculated corresponding to rejection (R_{obs}) of a solute having radius R:

$$R_{obs} = 1 - \frac{C_P}{C_0} = \frac{1}{(2\pi)^{0.5}} \int_{-\infty}^{U_R} \exp\left(-\frac{U^2}{2}\right) dU \qquad (5.23)$$

Having determined U_R, the parameters σ and m were estimated from the following fitting:

$$U_R = \frac{\ln R - m}{\sigma} \qquad (5.24)$$

5.2.5 Membrane Performance
5.2.5.1 Diffusive Permeability
Diffusive permeability (P_D) was measured using the experimental setup described in Figure 5.6, without any TMP drop, that is, in diffusion-governed mode. The samples were collected from the tanks at an interval of 15 minutes and measurements were made. P_D was calculated as[1-3]

$$P_D = \frac{\left\{\ln \frac{\Delta C_1(t_1)}{\Delta C_2(t_2)}\right\}}{A\left\{\dfrac{1}{V_a} + \dfrac{1}{V_b}\right\}(t_2 - t_1)} \qquad (5.25)$$

where

P_D is the diffusive permeability
$\Delta C_1(t_1)$ is the concentration difference between the feed and dialysate tanks at time (t_1)
$\Delta C_2(t_2)$ is the concentration difference between the feed and dialysate tanks at time (t_2)
A is the area of the membrane
V_a is the volume of feed
V_b is the volume of the dialysate

The diffusive permeability was calculated for urea, creatinine, vitamin B12, sucrose, and bovine serum albumin (BSA) and compared with available literature. The feed tank was filled with urea (600 mg/L), creatinine (20 mg/L), sucrose (50,000 mg/L), vitamin B12 (0.015 mg/L), and BSA (50 g/L) separately. Samples were collected from both feed and dialysate tanks, in an interval of 10 minutes. The urea concentration was determined by Ehrlich's reagent method,[4] and creatinine was measured using a standard creatinine kit. Vitamin B12 and BSA were measured using the UV–VIS spectrophotometer (supplied by M/s. Perkin Elmer, Connecticut, USA) at 361 and 280 nm, respectively. Sucrose was measured using a refractometer.

5.2.5.2 Clearance of Urea and Creatinine
Clearance[5] and its definition have been discussed in Chapter 2.

5.2.5.3 k_0A Values

k_0A[6-8] and its definition have been discussed in Chapter 2. k_0A is often used as an indicator, classifying the type of dialyzers, that is, high efficiency and high performance. Efficiency is associated with the passage of urea and creatinine, and "performance" or "flux" is associated with convective transport of water and/or middle-molecular-weight solutes, that is, β_2-micrglobulin (molecular weight 11,800 Da).[7] Various dialyzers in terms of k_0A values are available in literature.[8] k_0A of the prepared hollow fibers was measured.

5.2.5.4 Adequacy of Dialysis and Kt/V Values

Kt/V[9] has also been discussed in Chapter 2.

5.2.5.5 Effect of Feed and Dialysate Flow Rates

The transient behavior and the removal efficacy of uremic toxins of the membranes were studied by varying the relative flow rates of the feed and dialysate. The feed tank contained urea (600 mg/L) and subsequently creatinine (20 mg/L) dissolved in distilled water. The dialysate tank was filled with distilled water. The feed and dialysate flow rates were maintained at ratios of 1:1 and 1:2, at a scaled-down flow rate for the available area in a laboratory-prepared cartridge.

5.2.5.6 Scaling of Experiments

In order to test the performance of the spun membranes, it is important to scale the experiments. Only 300 fibers were used to fabricate the cartridge; hence, the volumetric flow rate that these fibers could handle was much less than that of an actual hemodialyzer. Thus, in order to replicate the values of flow rates maintained in a standard 1 m² cartridge (typically 250 mL/min), the Reynolds number in one fiber was determined:

$$\text{Re} = \frac{2\rho Q_B}{\mu n \pi R_i} \tag{5.26}$$

where
ρ is the density of the blood side solution
μ is its viscosity
n is the number of fibers
R_i is the inner radius of the fiber

Clearance (C_L) has been defined in Equation 2.4, and substituting it in Equation 5.26:

$$C_L = \Delta C^* \times \text{Re} \times \frac{\mu n \pi R_i}{2\rho} \tag{5.27}$$

where $\Delta C^* = \dfrac{C_{B_i} - C_{B_0}}{C_{B_i}}$.

Now, considering cases 1 and 2 for two different cartridges and keeping Re equal for both the laboratory-fabricated cartridge and the commercially available one having 7500 fibers, the scaling rule becomes

$$\frac{C_{L_1}}{C_{L_2}} = \frac{\Delta C_1^* n_1}{\Delta C_2^* n_2} \tag{5.28}$$

ΔC_1^* was measured experimentally with $n_1 = 300$. ΔC_2^* for a commercial cartridge, with flow rate 250 mL/min and 7500 (n_2) fibers (effective filtration area 1 m²), is 0.6.² Therefore, clearance (C_{L_2}) was calculated using Equation 5.28 for a cartridge with 1 m² filtration area and 250 mL/min for flow rate.

5.2.6 Membrane Spinning
5.2.6.1 Conventional Extrusion Mechanism
A simplified cross-sectional view of a spinneret is depicted in Figure 5.7a. It is evident that the polymers flow surrounding the water or anti-solvent. The water and polymers come out from the bore opening and when in contact with each other, mass transfer occurs as depicted in Figure 5.7b. The mechanism and principle of pore formation are described in previous chapters and the principle is the same in this case. The anti-solvent diffuses from the inner core to the surrounding polymer solution, and the solvent counterdiffuses toward the anti-solvent. This mechanism forms pores and macrovoids, typical characteristic features of a membrane. A delicate balance between the various operating conditions has to be achieved in order to extrude clinical grade hollow-fiber membranes.

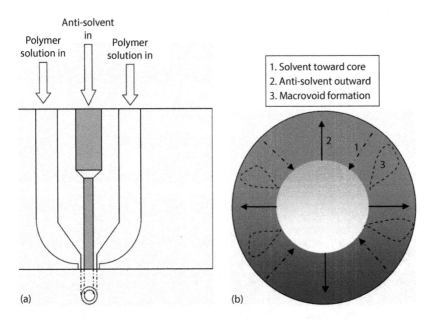

Figure 5.7 (a) Typical cross section of spinneret and (b) annular flow of polymer and anti-solvent.

5.2.6.2 SIS Assembly and Extrusion

It was first developed by Thakur and De[10] in 2012. It consisted of a larger-diameter outer syringe that had a hole drilled on its lateral surface. A smaller-diameter needle was bent at a particular angle and inserted through the hole, such that the two were concentrically arranged and the assembly was sealed to avoid leakage. The first step is to understand the basic device for spinning the fibers, which is depicted in Figure 5.8.

The polymer flows from the feed tank under pressure because of the use of a nitrogen cylinder. Water is pumped from the water tank. While water flows through the inner syringe of the SIS assembly (Figure 5.9), the polymer flows in the annular space surrounding it. At the outlet of the SIS, the polymer gets solidified in contact with water, forming the hollow core. Finally, the hollow fiber falls into the gelation bath containing water (Figure 5.8), thereby completing the phase inversion, and the fibers are then wound by the spool (Figure 5.8).

This was the commencing point of the present research effort. Initially, the SIS utilized a 16-gauge outer syringe (diameter 1600 μm) and a 24-gauge inner syringe (diameter 550 μm). However, the hollow-fiber membranes spun from this had a 400 μm inner diameter and 150 μm thickness. This is almost double the specifications for dialysis membranes. Hence, a series of SIS assemblies were fabricated and hollow-fiber membranes were spun varying the operating conditions. A summary of the most relevant efforts is presented in Figure 5.10. It was seen that an 18-gauge outer syringe (diameter 840 μm) and

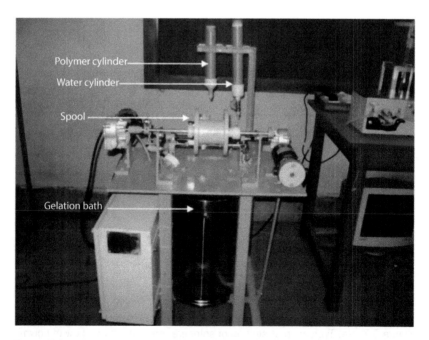

Figure 5.8   Low-cost spinning assembly developed in the lab.

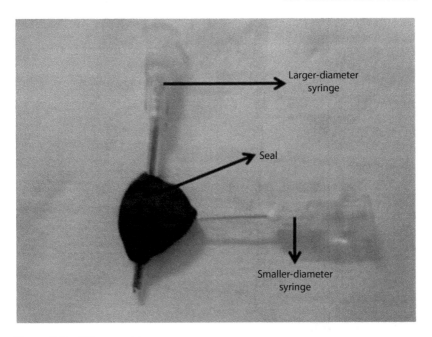

Figure 5.9   SIS assembly.

a U-100 insulin syringe (diameter 130 μm) yielded a hollow-fiber membrane of 240 μm but the thickness was too high (Figure 5.10a). However, this was an improvement with respect to the 400 μm fibers. As discussed, the water flow was facilitated through a pump till this juncture, but once smaller-diameter syringes were used, pump-driven water flow ruptured the walls of the hollow-fiber membranes due to flow instability. Hence, the pump was replaced by simple gravity-driven flow of water. This allowed smaller-gauge syringes, such as 19-gauge (diameter 690 μm) and U-100 (diameter 130 μm), to be used followed by 20-gauge (diameter 600 μm) and U-100 assemblies (diameter 100 μm) as depicted in Figure 5.10b and c, respectively. All the hollow-fiber membranes obtained were improved versions but did not adhere to the clinical specifications. After a few more iterations, a design was decided upon finally, which is depicted in Figures 5.10d and 5.11. As observed in Figure 5.10d, the 22-gauge (diameter 410 μm) and 32-gauge (diameter 110 μm) combination resulted in the desired specifications of hollow fibers. Figure 5.11 illustrates the final design, which helped in successful spinning of hemodialysis grade fibers.[11,12]

The final design consisted of a BD 32-gauge inner syringe (diameter 110 μm; Figure 5.11(IV)) that was bent at an angle of 120° and inserted into a 22-gauge outer syringe (diameter 410 μm). The assembly was then sealed. This is depicted in Figure 5.11(I) and (II). Figure 5.11(III) illustrates the arrangement of polymer- and water-holding tanks. The experimental setup consisted of a nitrogen cylinder that facilitates the flow of polymer solution. Water was

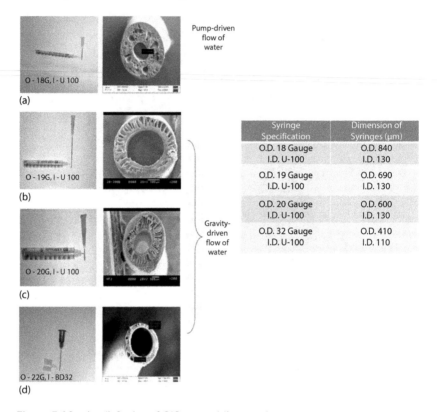

Figure 5.10   (a–d) Series of SIS assemblies used.

driven entirely by the gravitational head. The arrangement of the polymer- and water-holding tanks was spaced with the specific flange-to-flange and centerline distances as depicted in Figure 5.11(III). The SIS assembly was fitted into the clouded portion of the assembly. Finally, a look at Figure 5.11(IV) shows the overall design of the unit. The water head to be maintained for the spinning was 160 cm from the ground level. The cost of an SIS assembly and the spinning unit was $0.3 and $1800, respectively. The spinning conditions are given in Table 5.1.

5.2.7 CFD Simulation of Flow to Fine-Tune the Design of Extrusion Assembly

The basic difference between a spinneret and an SIS assembly is the coaxial flow arrangement. In typical spinnerets, the flow of polymer and that of water are coaxial. In SIS, the bent part of the inner syringe (inserted into the larger syringe) creates an obstruction in the path of polymer flow. This may destabilize the flow patterns and velocity distributions causing disruption in fiber formation. Hence, in order to understand the extent of the influence of the bent part and investigate the propagation of disturbance, a CFD analysis has been carried out.

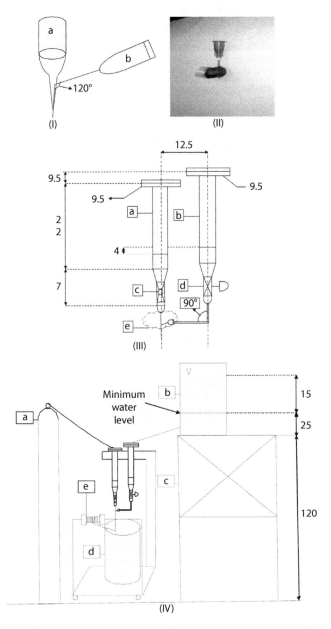

Figure 5.11 (I) Syringe assembly, (a) disposable syringe, (b) insulin syringe. (II) Image of the syringe assembly. (III) Detailed drawing of the polymer and water cylinders: (a) polymer tank, (b) water tank, (c) ball valve, (d) needle valve, (e) syringe assembly. (IV) Schematic diagram of the spinning setup: (a) nitrogen cylinder for pressurizing, (b) bucket, (c) steel structure, (d) gelation bath, and (e) spool for winding the fibers. All dimensions are in centimeters.

Table 5.1 Conditions for Spinning of Dialysis Fibers	
Parameters	**Values**
Pressure	140–200 kPa
Water flow rate	0.2 mL/min
Polymer flow rate	2 mL/s
Air gap	25 cm
Minimum water head	165 cm
Winding speed	2 cm/s
Residence time in gelation bath	30 s
Outer diameter of outer syringe (Dispovan-22G)	700 μm[a]
Inner diameter of outer syringe (Dispovan-22G)	410 μm[a]
Outer diameter of inner syringe (BD-32G)	230 μm[a]
Inner diameter of inner syringe (BD-32G)	110 μm[a]

[a] Sigma Aldrich website.

The geometry was drawn in GAMBIT and was imported and solved using FLUENT. The scheme for meshing was Tet/Hybrid, and the grid type chosen was a T grid. The mesh node spacing was 0.02 with 169,179 elements. The total number of nodes was 33,517 and was solved with the following equations and boundary conditions:

Continuity equation:

$$\nabla \cdot \vec{v} = 0 \tag{5.29}$$

and general momentum conservation equation:

$$\frac{\partial}{\partial t}\left(\rho \vec{v}\right) + \nabla \cdot \left(\rho \vec{v}\vec{v}\right) = -\nabla p + \nabla \cdot \left(\bar{\bar{\zeta}}\right) + \rho \vec{g} \tag{5.30}$$

where

ρ is the density of the polymer solution
$\vec{v}$ is the velocity field
p is the pressure
$\left(\bar{\bar{\zeta}}\right)$ is the stress field tensor, considering power law rheology of polymer
solution ($n = 0.98$; $k = 3.3$ Pa s$^{0.98}$)

The relevant boundary conditions were inlet pressure, 140 kPa, and outlet pressure, 0 kPa. No slip boundary conditions were used on the inner and outer walls.

5.3 Results and Discussion

5.3.1 FTIR Analysis

The polysulfone (PSf) chain is identified (Figure 5.12a) by the major character-istic adsorption band of sulfonate ($-SO_2^-$) groups (1322–1323 cm⁻¹ and 1293–1294 cm⁻¹) and the C–O–C stretching mode (1241–1244 cm⁻¹).[13] In addition, an aromatic >C=C< stretching vibration at 1635 cm⁻¹ is also observed.[14] The peak at 3336 cm⁻¹ is attributed to stretching vibrations of the O–H groups of the adsorbed water molecules (Figure 5.12a).

5.3.1.1 Interaction in Solid Film

5.3.1.1.1 Interaction of PSf and PEG (PVP 0)

The observable peaks (Figure 5.12b) are 3407 cm⁻¹ (strong), which are attrib-uted to terminal O–H stretching vibrations of polyethylene glycol (PEG), and 1654 cm⁻¹, due to the aromatic >C=C< absorption band of PSf. C–O–H bending is observed at 1390 cm⁻¹.[15] Another peak at 1254 cm⁻¹ is attributed to C–O–C stretching.[4] A characteristic peak at 1100 cm⁻¹ is also observed for C–O stretching of PEG.[16] However, the shift in C–O–C stretching frequency from 1244 to 1254 cm⁻¹ indicates the chemical interaction between PSf and PEG.[17] The summary of the interactions in film is presented in Table 5.2. In FTIR analysis, a blue shift indicates the shifting of a band to higher wave number

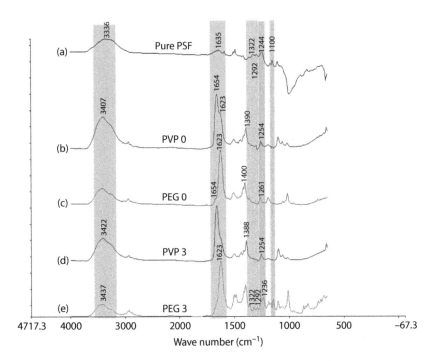

Figure 5.12　All solid (film) phase.

109

Table 5.2 Functional Groups and Stretching Frequencies of the Blend in Film

Functional Groups	−O–H (cm⁻¹)	>C=O (cm⁻¹)	>C=C< (cm⁻¹)	C–O–H/O–C–N (cm⁻¹)	SO₂ (cm⁻¹)	C–O–C (cm⁻¹)	C–O (cm⁻¹)
Case							
Pure PSf	3336	—	1635	—	1322, 1292	1241–1244	—
Pure PVP	—	1645 [17]	—	1385 [15]	—	—	—
Pure PEG	3335 [14]	—	—	—	—	—	—
In film							
PVP 0	3407 (blue shift)	—	1654	1390	—	1254 (blue shift)	1100
PEG 0	—	1654 (blue shift)	—	1400	—	1261 (blue shift)	1100
PVP 3	3422 (blue shift)	—	—	1388	—	1254	1100
PEG 3	3437 (blue shift)	—	—	—	1322, 1293	1236 (red shift)	1100

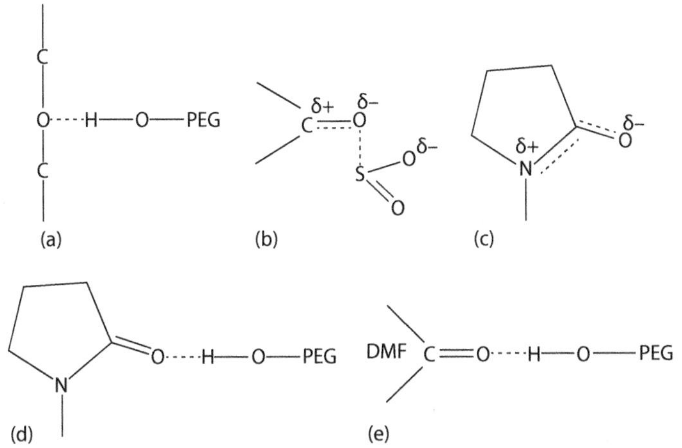

Scheme 5.1 (a) PSf–PEG interaction, (b) PVP–PSf dipole–dipole interaction, (c) O–C–N deformation of PVP, (d) PVP–PEG interaction, and (e) PVP–DMF interaction.

regions due to interactions such as hydrogen bonding and dipole–dipole interaction. The physical significance of this is strengthening of a particular bond. It may be noted that a blue shift of the terminal –O–H– stretching of PEG from 3335 to 3407 cm^{-1} can be attributed to strong interaction with PSf. This is presented in Scheme 5.1a, which illustrates the PSf–PEG interaction.[14] The characteristic absorption band for sulfonate group (–SO$_2^-$) of PSf disappears in the polymer blend.

5.3.1.1.2 Interaction of PSf and PVP (PEG 0)

With the addition of PVP (Figure 5.12c), the interaction consists mainly of the dipole–dipole interaction of the >C=O group in the PVP chain and the sulfonate group in the PSF chain. The PVP–PSf dipole–dipole interaction is presented in Scheme 5.1b. The >C=O of PVP is shifted from 1645 to 1654 cm^{-1}.[17] The peak is of low intensity due to less amount of PVP. Additionally, one peak at 1400 cm^{-1} is attributed to O–C–N deformation. This is depicted in Scheme 5.1c as O–C–N deformation of PVP[18] and another one at 1261 cm^{-1} attributed to C–O–C stretching of PSf. Changes in stretching frequency from 1244 to 1261 cm^{-1} also indicate the interaction between PSF and PVP (Scheme 5.1b and Table 5.2). The higher shift in stretching frequency indicates comparatively stronger interaction of PSF–PVP (Scheme 5.1b) than of PSf–PEG (Scheme 5.1a). It can be mentioned that the characteristic –SO$_2$– stretch (a twin at 1322 and 1293 cm^{-1}) is also not observed (Table 5.2) and may be shifted somewhere else due to interpolymeric interactions.

5.3.1.1.3 Addition of PVP into PSf–PEG

Addition of PVP to the PSf–PEG blend displays peaks at 3422, 1388, and 1254 cm^{-1}, owing to –O–H– stretching (of PEG), –O–C–N– deformation (of PVP), and –C–O–C– stretching (of PSf), respectively (Figure 5.12d).

Additionally, a blue shift, or strengthening of bond, of –O–H– stretching frequency from 3407 to 3422 cm^{-1} is observed (Table 5.2), which may be attributed to the interpolymeric interaction between PVP and PEG (depicted in Scheme 5.1d).

5.3.1.1.4 Addition of PEG into PSf–PVP

However, with the addition of PEG into the PSf–PVP matrix, there is a blue shift in O–H stretching (3407–3437 cm^{-1}) and a red shift in the C–O–C stretch (1261–1236 cm^{-1}) (interaction between PVP and DMF is presented in Scheme 5.1e). This is visible from Figure 5.12e. The red shift is the opposite of the blue shift and indicates weakening of bonds. This indicates weakening of PSf and PEG interaction and strengthening of interaction between PVP and PEG (Scheme 5.1a). It is also observed that appearance of the characteristic –SO$_2$– stretch (a twin at 1322 and 1293 cm^{-1}) of PSf indicates that it is relatively free from interactions with PVP and PEG.

5.3.1.2 Interaction in Solution
5.3.1.2.1 PSf–PEG Interaction in DMF

In DMF solution, the major characteristic peaks of the polymers are at 3483 cm^{-1} attributed to O–H stretching and at 1385 cm^{-1} attributed to C–O–H bending of PEG (Figure 5.13). The peaks at 1671–1646 cm^{-1} split into three

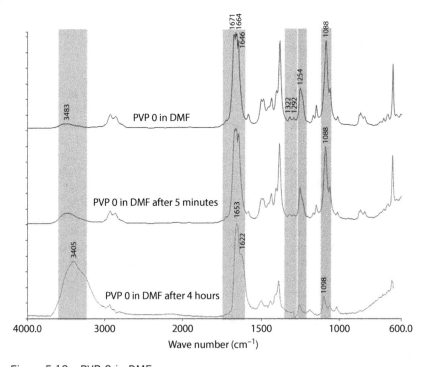

Figure 5.13   PVP 0 in DMF.

peaks of 1671, 1664, and 1646 cm^{-1}, and these are attributed to amide of DMF. The rest of the two are for aromatic >C=C< bonds. The two absorption bands are due to two distinct and significant conformers of the polymer. After complete evaporation of the DMF, the peaks are resolved into two peaks at 1653 and 1622 cm^{-1}, characteristic of C=C of PSf. The characteristic –SO$_2$– stretch (a twin at 1322 and 1292 cm^{-1}) is clearly visible in the solution phase, which disappears completely after evaporation of DMF, indicating weak interaction between PSf and solvent DMF. In addition, there is a red shift of the O–H stretching frequency from 3483 to 3405 cm^{-1} and a blue shift of C–O from 1088 to 1098 cm^{-1} (Table 5.3 summarizes the interactions in solution phase). The red shift of O–H occurs after complete evaporation of DMF, indicating that PEG and DMF interaction (Scheme 5.1e) is strong and the blue shift of the C–O stretch indicates weak interaction between PSf and DMF.

5.3.1.2.2 PSf–PVP Interaction in DMF

In the solution phase of the PSf–PVP blend, distinct peaks at 1640, 1628, 1395, and 1100 cm^{-1} (Figure 5.14) are observed. The peak at 1640 cm^{-1} is due to amide of DMF; the 1628 cm^{-1} band indicates aromatic >C=C< absorption in PSf, the 1395 cm^{-1} peak depicts O–C–N deformation in PVP, and the 1100 cm^{-1} peak depicts the C–O vibration of PSf.[19] After complete evaporation of DMF, the twin peaks at 1640 and 1628 cm^{-1} merge to 1622 cm^{-1} and the peak at 1395 cm^{-1} is shifted to 1400 cm^{-1} indicating weak dipole–dipole interaction between DMF and PVP. The appearance of the characteristic –SO$_2$– stretch at 1322 and 1292 cm^{-1} in solution indicates no interaction between DMF and PSf. Due to strong interaction between PSf and PVP, the peaks discussed completely disappeared or shifted somewhere after evaporation of DMF.

5.3.1.2.3 Interaction of PVP with the PSf–PEG Blend in DMF

After addition of PVP with the PSf–PEG blend in DMF solution, major characteristic absorption bands at 3492, 1385, and 1248 cm^{-1} are observed. This is attributed to O–H stretching of PEG, C–O–H deformation of PEG, and C–O–C stretching mode of PSf, respectively (Figure 5.15). The peak at 1671 cm^{-1} is due to the amide of DMF, whereas the peaks at 1661 and 1646 cm^{-1} are due to two aromatic >C=C< absorptions. After complete evaporation of DMF, there is a red shift from 3483 to 3405 cm^{-1} and two blue shifts from 1385 to 1390 cm^{-1} and 1090 to 1098 cm^{-1}. The strong red shift in O–H indicates weakening of O–H bond of PEG due to strong interaction between PEG and DMF (Scheme 5.1e). At the same time, the blue shift in C–O–H bending of PEG and the C–O–C stretching mode of PSf indicates the strong interaction between PSf and PEG. Addition of PVP does not interfere with the interaction of PSF and PEG. The characteristic –SO$_2$– stretching mode completely disappears after phase inversion.

5.3.1.2.4 Interaction of PEG with PSf–PVP Blend in DMF

With the addition of PEG to PSf–PVP, the blend displays peaks at 3488, 1640, 1628, and 1392 cm^{-1}. The 3488 cm^{-1} peak is due to the presence of O–H stretching in PEG, the 1640 cm^{-1} band is due to amide of DMF, the 1628 cm^{-1}

Table 5.3	Functional Groups and Stretching Frequencies of the Blend in Solution						
Case / **Functional Groups**	–O–H (cm⁻¹)	>C=O (cm⁻¹)	>C=C< (cm⁻¹)	C–O–H/O–C–N (cm⁻¹)	SO₂ (cm⁻¹)	C–O–C (cm⁻¹)	C–O (cm⁻¹)
Case							
Pure PSf	3336	—	1635	—	1322, 1292	1241–1244	—
Pure PVP	—	1645 [17]	—	1385 [15]	—	—	—
Pure PEG	3335 [14]	—	—	—	—	—	—
In solution							
PVP 0	3405 (red shift)	—	1664, 1646	1385	1322, 1292	—	1098 (blue shift)
PEG 0	—	—	1628	1395	1322, 1292	—	1100
PVP 3	3492	—	1661, 1646	1385	1322, 1292	1248	—
PEG 3	3488 (red shift)	—	1628	1392 (blue shift)	1322, 1292	—	—

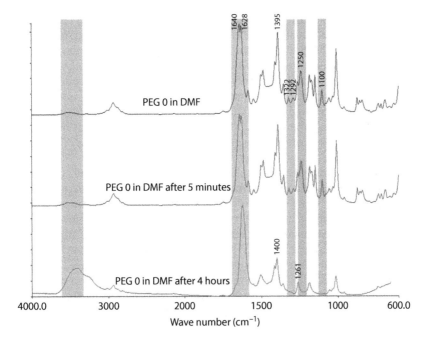

Figure 5.14 PEG 0 in DMF.

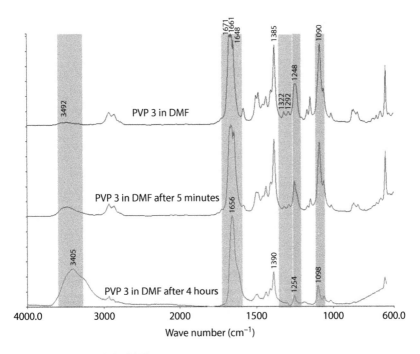

Figure 5.15 PVP 3 in DMF.

peak is due to aromatic >C=C< absorption in PSf, and the 1392 cm⁻¹ band is due to O–C–N deformation in PVP (Figure 5.16). There was a red shift of O–H stretching from 3488 to 3410 cm⁻¹, indicating interaction of PEG and DMF (Table 5.3). The blue shift from 1392 to 1398 cm⁻¹ is observed due to strong interaction of >C=O of PVP and O–H of PEG (Scheme 5.1d and Table 5.3). The >C=O stretch of PVP does not appear due to overlapping with >C=C< absorption. Interestingly, the characteristic –SO₂- stretch of PSf at 1322 and 1292 cm⁻¹, which did not appear after phase inversion in the previous case (in film), clearly appears after phase inversion in this particular case. This may be due to strong interaction between PVP and PEG.

It can be summarized from the aforementioned observations that in the solution phase, the interpolymeric interactions are poor. PSf interaction with the solvent (DMF) is weak. However, PVP and PEG due to the presence of strong polar groups (such as >C=O, –OH) interact strongly with the polar part of the solvent, causing either a "blue" or a "red" shift of the bands in their ATR–FTIR spectra. However, after complete phase inversion, the stronger interaction of PSf–PVP compared to PSf–PEG is also noticed. Addition of PVP to the PSf–PEG blend leads to significant interaction between PVP and PEG, weakening the interaction with PSf. This causes the appearance of characteristic –SO₂- stretching frequency of PSf. With this understanding of interactions between polymers, it is important to see the effect in rheology of the blend.

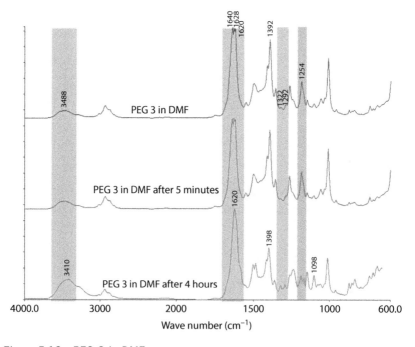

Figure 5.16 PEG 3 in DMF.

5.3.2 Rheology Analysis

The model chosen for analyzing the polymer blend solution is power law fluid. The plots of shear stress to shear rate are depicted in Figure 5.17. From this figure, the values of flow behavior index (n) and consistency index (k) are obtained. In the case of PEG variation, an average k can be assumed as 17.6 Pa s$^{1.27}$ and average n as 1.27. For PVP variation, the same was 3.3 Pa s$^{0.98}$ and 0.98, respectively.

The double logarithmic plots of storage (G') and loss modulus (G'') are depicted in Figure 5.18. It is seen that at lower frequencies the loss modulus is less than the storage modulus. This signifies that at such frequencies the blend

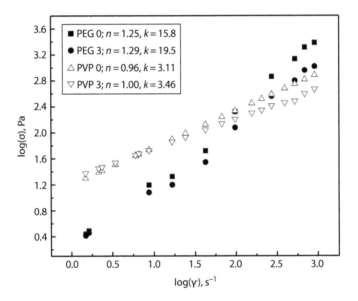

Figure 5.17   Double logarithmic plot of shear stress versus shear rate.

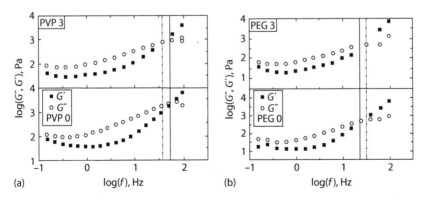

Figure 5.18   Double logarithmic plot of shear moduli (G' and G'') versus oscillatory frequency (a) PVP addition and (b) PEG addition.

117

behaves like a viscous liquid with irrecoverable viscous losses. At higher frequencies, the storage modulus becomes higher than the loss modulus, indicating dominance of solid-like characteristics.[20] This also signifies reversibility of energy storage in the sample. G' and G'' vary as $f^{0.47-0.58}$ and $f^{0.75}$, respectively, for PVP variation. G' and G'' for PEG variation vary with frequency as $f^{0.47-0.57}$ and $f^{0.79-0.98}$, respectively. The exponent of f decreases with addition of PVP or PEG, signifying interaction within the polymeric chains.[20] The important aspect to understand here is the shift of crossover point ($G' = G''$). With increasing addition of PVP the crossover frequency shifts left, whereas with increasing addition of PEG the same shifts right. This can be attributed to the interactions as discussed for the FTIR analysis. The polymeric chain interactions in case of PVP addition are such that the transition from liquid-like to solid-like behavior is attained faster with concentration of PVP. This is just the opposite in case of PEG addition, where the transition from liquid to solid is delayed with PEG concentration. As discussed in the section on FTIR analysis, the PEG and PVP chains interact more with PEG concentration, as compared to the reverse case. Thus, the oscillatory shear required to break those interactions is more in this case.

The double logarithmic plot of loss factor (tan $\delta = G'/G''$) versus frequency is illustrated in Figure 5.19. The loss factor indicates the relative amount of energy dissipated by a material during cyclic stress.[20] The free volume decreases once there are favorable interactions between the polymer chains and this increases the elastic behavior, thereby decreasing the loss factor. In the figure, it is clear that the loss factor in case of PEG variation is lesser than that of PVP variation. This, again, is due to the polymer chain interactions and hence an increase in the elastic behavior of the blend.

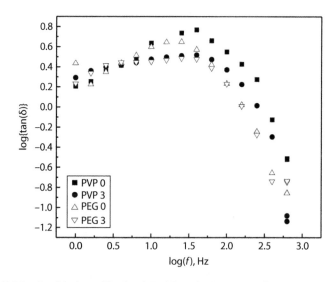

Figure 5.19   Double logarithmic plot of loss factor versus frequency.

These results are significant with respect to membrane engineering. The overall summary of the FTIR analysis and rheology study are

1. PEG variation in the presence of PSf and PVP increases the interaction of PVP and PEG chains. This interaction has an impact on the rheological behavior.

2. The influence of increasing interaction is felt when the shift in the crossover frequency is observed. It is evident that increasing PEG in the blend induces more elastic behavior. This is also observed in the loss factor curves, where the PEG variation curves depict lower loss factor values.

3. In case of membrane engineering, it can be concluded that increasing PEG in a blend would induce elastic behavior, whereas increasing PVP would do just the opposite. Keeping in mind the basics of membrane engineering (as also discussed in Chapter 4), inducing more PVP would quicken the demixing process, whereas increasing PEG would delay it, attributed to elastic–viscous properties. Hence, PVP addition would increase the porosity of the membranes and is discussed in the rest of the chapter.

5.3.3 MWCO and the Surface Morphology of Membranes

In light of the conclusions drawn earlier, it is easier to understand membrane characteristics. Figure 5.20 depicts the MWCO of the three membranes. S1 has an MWCO of 6 kDa, whereas for S2 and S3 the values are 12 and 16 kDa, respectively. These results corroborate with the discussion in the preceding section. Moreover, PVP not only acts as "pore former," it also increases the hydrophilicity of the blend. The addition of PVP leading to the formation of

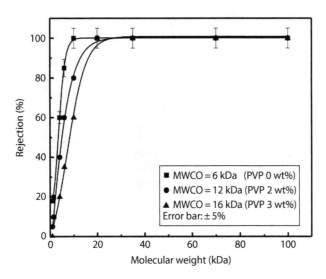

Figure 5.20 MWCO of the dialysis grade membranes.

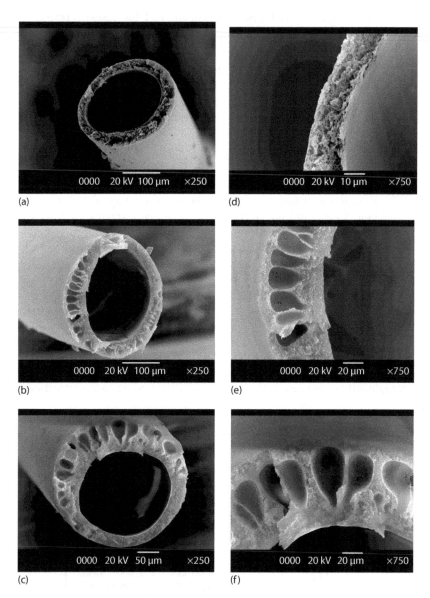

Figure 5.21 SEM images of the dialysis membranes: cross section (a) 6 kDa, no PVP; (b) 12 kDa, 2% PVP; (c) 16 kDa, 3% PVP; thickness (d) 6 kDa, no PVP; (e) 12 kDa, 2% PVP; and (f) 16 kDa, 3% PVP.

pores is shown in SEM images in Figure 5.21. It is seen that S1 has a dense structure, but S2 and S3 have more porous structures. There is a dense skin layer followed by a porous substructure for S2 and S3, and the porous nature is more pronounced for S3 than for S2 owing to higher PVP concentration.

5.3.4 Permeability and Contact Angle Results

Figure 5.22 depicts the permeability and contact angle with PVP concentration. It is observed that the lower the contact angle, the higher the permeability of the membranes. This is attributed to the interplay of two aspects. First, membrane hydrophilicity increases with PVP concentration. Second, as discussed, higher PVP induces more porous structure enhancing the transport capabilities of membranes. This is evident from Figure 5.22, as S1 has the highest contact angle of 75° with the lowest permeability of 2×10^{-8} m^3/m^2 s mmHg, whereas S3 has the lowest contact angle of 64° with the highest permeability of 5×10^{-8} m^3/m^2 s mmHg. S2 lies in between S3 and S1.

5.3.5 Porosity, Pore Size Distribution, and Tensile Strength

Figure 5.23a depicts the porosity and tensile strength of the membranes, and Figure 5.23b predicts the pore size distribution of the membranes. S1, S3, and S3 membranes have porosity 21%, 28%, and 31%, respectively. This porous structure decreases the mechanical strength of the membranes, consequently decreasing the breaking stress. Thus, as PVP concentration is increased from 0 to 3 wt%, the breaking stress is reduced from 13.7 to 8.7 MPa. The pore size distribution (Figure 5.23b) follows the same pattern. As the PVP concentration increases from 0 to 2 wt%, membranes become more porous and the mean pore radius increases from 1 to 2.5 nm. It is reported in the literature[21] that the high-performance membrane (HPM) synthesized by Rhone Poulenc

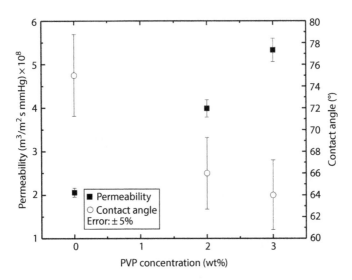

Figure 5.22 Permeability and contact angle as a function of PVP concentration.

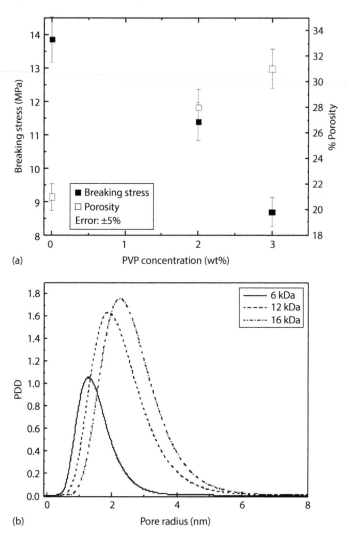

(a)

(b)

Figure 5.23 (a) Tensile strength and porosity as a function of PVP concentration. (b) Pore size distribution of the membranes.

(AN 69) has a pore size of 29 Å, whereas polyamide membranes (Gambro) have externally macroporous pore sizes of 40–50 Å. The 6 kDa membrane has an average pore size of 30 Å, whereas the 12 and 16 kDa membranes in this study have pore sizes of 40–42 Å, as seen in Figure 5.23b.

5.3.6 Diffusive Permeability

The diffusive permeability values are presented in Table 5.4. It is observed that the membranes exhibit comparable P_D values with respect to the literature. This parameter indicates the ease of diffusion of various solutes

Table 5.4	Diffusive Permeabilities of the Membranes			
Components	Reported Values[38] ($\times 10^4$), cm/s	6 kDa ($\times 10^4$), cm/s	12 kDa ($\times 10^4$), cm/s	16 kDa ($\times 10^4$), cm/s
Urea	15.2	14.1 ± 1.1	15.4 ± 1.2	16 ± 1.2
Creatinine	8.8	7.4 ± 0.6	8.5 ± 0.07	9.1 ± 0.07
Sucrose	2.4	1.9 ± 0.01	2.3 ± 0.11	2.5 ± 0.11
Vitamin B12	0.25	0.22 ± 0.001	0.25 ± 0.001	0.25 ± 0.001
BSA	0	0	0	0

across the membranes. As the molecular size of the solutes increases, their diffusion coefficient decreases. Urea has a molecular weight of 60 Da and its diffusive permeability for a 6 kDa membrane is 14.1×10^{-4} cm/s, 15.4×10^{-4} cm/s for a 12 kDa membrane, and 16×10^{-4} cm/s for a 16 kDa membrane. Creatinine has a molecular weight of 113 Da and has a diffusive permeability of 7.4×10^{-4} cm/s, 8.5×10^{-4} cm/s, and 9.1×10^{-4} cm/s for the 6, 12, and 16 kDa membranes, respectively. Hence, it is observed that doubling the molecular weight reduces the diffusive permeability by about 42%. On the other hand, sucrose has a molecular weight of 342 Da, and its diffusive permeability is 1.9×10^{-4} cm/s, 2.3×10^{-4} cm/s, and 2.5×10^{-4} cm/s for the 6, 12, and 16 kDa membranes, respectively. Sucrose has a molecular weight almost six times that of urea resulting in decrease in P_D value by 90%. Similarly, vitamin B12 has a molecular weight of 1355 Da and hence has a diffusive permeability varying from 0.22×10^{-4} to 0.25×10^{-4} cm/s, which is a 98% decrease with respect to urea. This observation is expected as larger-sized molecules diffuse slowly through membranes. BSA has a molecular weight of 66 kDa and, expectedly, has a diffusive permeability of zero due to size exclusion.

5.3.7 Urea and Creatinine Clearances
From clearance values (Figure 5.24a through c), it can be inferred that the 12 and 16 kDa membranes are potential HPMs and a 6 kDa membrane is a conventional high-efficiency dialysis membrane. Figure 5.24 shows that clearance increases with feed flow rate and the highest clearance is realized at 300 mL/min. Urea and creatinine clearances at this flow rate are 200 mL/min for 12 and 16 kDa and 180 mL/min for 6 kDa. As expected, creatinine clearance is slightly lower than urea clearance due to its larger molecular weight. Creatinine clearances at 300 mL/min of feed flow rate are around 165–180 mL/min for the 12 and 16 kDa membranes and 150 mL/min for the 6 kDa fibers. In the reported literature, clearances of high-performance dialyzers are in range of 200–300 mL/min. However, this

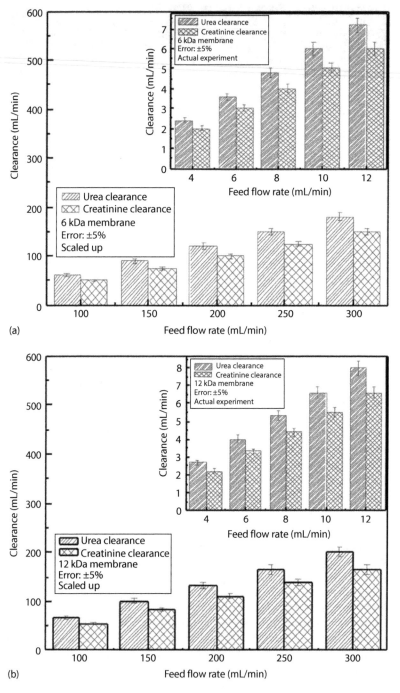

Figure 5.24 Clearance of urea and creatinine for the three membranes: (a) 6 kDa; (b) 12 kDa. (*Continued*)

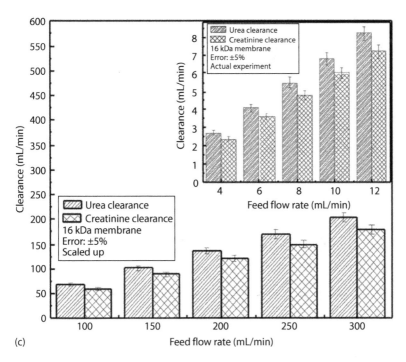

Figure 5.24 (*Continued*) Clearance of urea and creatinine for the three membranes: (c) 16 kDa.

is based on two aspects: (1) optimum design of dialyzers, that is, packing efficiency, inclusion of spacers and, most importantly, (2) convective mode of transport. These two effects make the transport of uremic toxins higher than what is reported in the present work and lead to clearance values less than what are reported.

5.3.8 k_0A and Kt/V

Figure 5.25 presents the effect of feed flow rate on k_0A of the membrane. The overall mass transfer coefficient of the membrane increases with the feed flow rate. Within the normal operating condition range (feed flow rate between 250 and 300 mL/min), it is observed that a 6 kDa membrane lies in the low-flux, low-efficiency range[5–8]. However, the k_0A values for the 12 and 16 kDa membranes are far higher, that is, around 450–475 mL/min. This places them in the high-flux, low-efficiency category.[5] However, since the clearance values discussed in this chapter are solely due to diffusion, the obtained values are underestimated. As discussed earlier, 12 and 16 kDa membranes have the potential to serve as HPM.

Kt/V of the membranes is presented in Figure 5.26. A patient weighing 70 kg requires approximately 350 minutes to reduce the urea level below 400 mg/L with the 6 kDa membrane, 300 minutes using the 12 kDa membrane, and

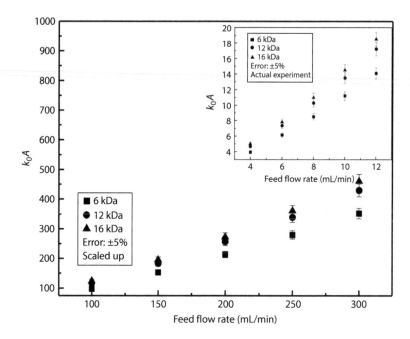

Figure 5.25 k_0A values for three membranes.

280 minutes using the 16 kDa membrane. The trend is similar for all the subsequent body weights of 60, 50, and 40 kg. Apparently, the time required is slightly on the higher side, but again this is due to the fact that the clearances are computed considering only diffusion.

5.3.9 Effect of Feed and Dialysate Flow Rates

As observed from Figures 5.27 and 5.28, urea concentration decreases in the order of 16 kDa > 12 kDa > 6 kDa. As the dialysate flow rate (DFR) is increased to twice of feed flow rate (FFR), the decline is faster due to enhanced driving force of mass transfer. It is observed that for FFR:DFR = 1:1, a 6 kDa membrane needs 250 minutes to reach the recommended concentration (400 mg/L), whereas a 12 kDa membrane needs 175 minutes and a 16 kDa one needs around 120 minutes. However, when FFR:DFR is 1:2, this time drops to 200 minutes for 6 kDa and 120 minutes for both 12 and 16 kDa. It can be inferred that there is marked improvement as far as performance of the 16 kDa membrane is concerned over the 6 kDa membrane. However, the 12 and 16 kDa membranes have comparable performance. Similarly, for creatinine, Figure 5.27 reveals that for an FFR:DFR ratio of 1:1 the 6 kDa membrane takes 300 minutes to reach the 12 mg/L level, whereas the 12 and 16 kDa membranes take 250 minutes to reach the level. However, for FFR:DFR 1:2, the fall is more rapid and the desired levels are reached within 175–250 minutes.

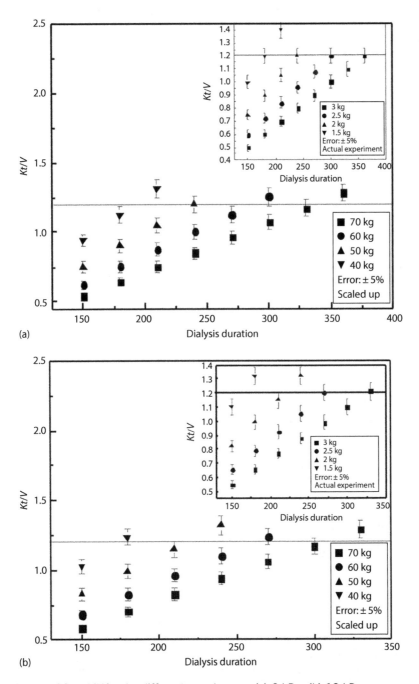

Figure 5.26   *Kt/V* for the different membranes: (a) 6 kDa; (b) 12 kDa.

(*Continued*)

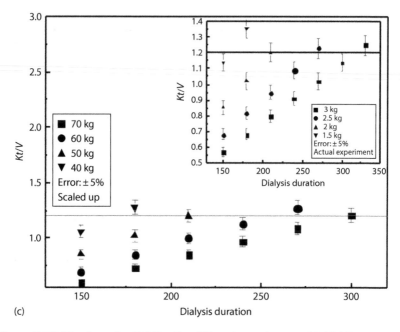

(c)

Dialysis duration

Figure 5.26 (*Continued*) *Kt*/*V* for the different membranes: (c) 16 kDa.

In the present work, experiments were conducted in purely the diffusion-governed mode. In addition, the concentration gradient, that is, the driving force across the membrane decreases with time as both the feed and the dialysate are recycled. Therefore, the dialyzer in this work performs poorly compared to the continuous supply of fresh dialysate in actual dialysis. However, even in this mode of operation, the dialyzer with low-cost spinning technology is able to reduce the level of uremic toxins to the desired limit in adequate time.

5.3.10 CFD Analysis Results
5.3.10.1 CFD Simulation of the SIS Assembly

Finally, ending the discussion with the design of the spinneret would be the most logical conclusion to this chapter. Hence, it will be interesting to investigate an optimum design of the SIS assembly. As depicted in Figure 5.7a, conventional spinnerets have an annular arrangement, but the SIS assembly has an obstruction in the flow path of the polymer, in the form of the bent portion of the needle (Figure 5.11(I)). This causes a disturbance in flow patterns evolving around the bent part and the effect of that obstruction propagates along the flow path. This is depicted in Figure 5.29, where the length of the bent part of the inner needle to the outlet is depicted as L. An optimum L should be maintained so that the disturbance in the flow pattern at the outlet of the

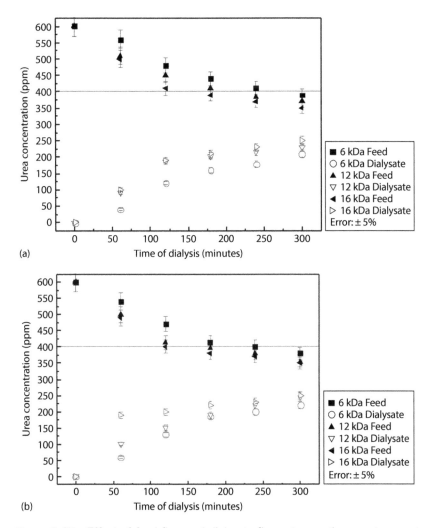

Figure 5.27   Effect of feed flow and dialysate flow rates on the urea transport through the membranes: (a) FFR:DFR = 1:1 and (b) FFR:DFR = 1:2.

needle is not felt. If this be the case, then nonuniform flow of the polymer solution would be realized, which is not suitable for spinning hollow-fiber membranes with uniform characteristics. This problem is addressed with CFD simulation.

The problem is worked out for various values of L, starting from 1 mm and ending with 6 mm. In the present design (as described in Figure 5.10), the value of L is 2 mm. However, the exercise is carried out in order to investigate the propagation of disturbance, if any, and the optimum L that

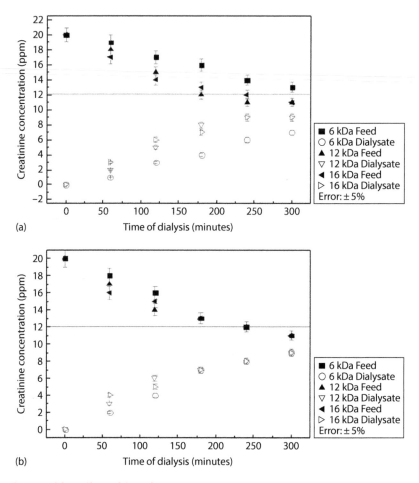

(a)

(b)

Figure 5.28   Effect of feed flow and dialysate flow rates on the creatinine transport through the membranes: (a) FFR:DFR = 1:1 and (b) FFR:DFR = 1:2.

is required to be maintained. GAMBIT is used to draw the basic geometry as depicted in Figure 5.29, with proper dimensions as in Table 5.1. The solution for velocity and shear stress profiles is illustrated in Figures 5.30 and 5.31 for two values of L (1 and 6 mm). It is evident that as the polymer solution approaches the bend, there is separation of flow, and this disturbance propagates as the solution continues to flow. However, the interesting aspect to note is that the disturbance in wall shear dies down very quickly, and at $y = 0.06$ mm down the bend, its distribution is uniform. On the other hand, the velocity distribution shows that the propagation in disturbance persists even at $y = 0.06$ mm, which attains a steady value at around $y = 1.0$ mm. At $L = 1$ mm, there is no sign of propagation of disturbance. Similar is

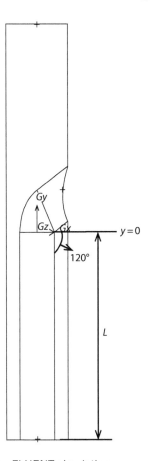

Figure 5.29 Geometry for FLUENT simulation.

the case for $L = 6$ mm. The disturbance does not propagate much and there is a uniform velocity distribution at the outlet. The reason for this exercise is illustrated in Figure 5.32. It is seen that as L is increased, the wall shear decreases and so does the outlet velocity. Reduction in the outlet velocity is not desirable as this would decrease the throughput of the process. On the other hand, minimizing wall shear is desirable as this reduces chances of shear-induced demixing occurring in those regions. Hence, an optimum distance has to be reached where throughput and wall shear can reach a compromise. This value is around 2.5 mm. Now, since it is observed that uniform velocity distribution is already reached at around $L = 1$ mm, maintaining $L = 2.5$ mm would ensure (1) uniform velocity distribution in the annular region, (2) minimum wall shear, and (3) maximum throughput at the outlet. In the present design (as described in Figure 5.11), L is maintained at 2 mm.

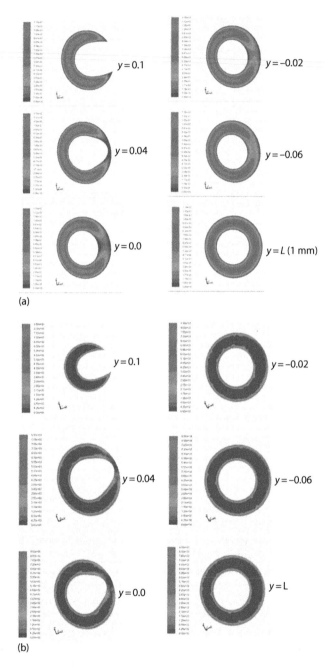

Figure 5.30 (a) Velocity profile for $L = 1$ mm and (b) shear stress profile for $L = 1$ mm.

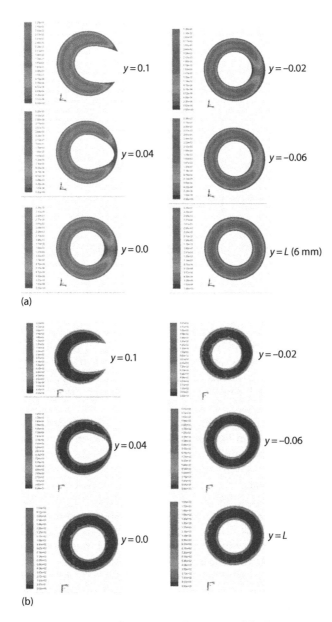

Figure 5.31 (a) Velocity profile for $L = 6$ mm and (b) shear stress profile for $L = 6$ mm.

133

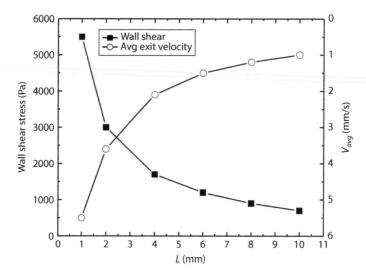

Figure 5.32 Variation of wall shear stress and average velocity of polymer solution with length of extrusion assembly.

5.4 Conclusion

In the present effort, a complete design of extrusion of hollow-fiber membranes, conforming to clinical specifications, was carried out. The problem was tackled starting from the basics of polymer interactions at the fundamental level. These interactions manifest in the form of varying properties in the rheological behavior of blend. It is observed that PEG 200 acts as a pore constrictor by inducing elastic behavior in a polymer blend. On the other hand, PVP acts reversely and helps in pore formation. Next, an in-house, low-cost SIS assembly was fabricated with a complete process design. This SIS assembly costs only $0.30 and is used to spin hemodialysis grade hollow-fiber membranes with 220 μm inner diameter and 35–40 μm thickness. Various grades of dialysis membranes are spun varying the PVP concentrations for synthesizing high-efficiency and high-performance dialysis membranes. These are subjected to in vitro performance testing and clinical parameter estimations. It was observed that the spun membranes had diffusive permeability values closely matching those reported in the literature. Moreover, these membranes transported uremic toxins appreciably in the diffusion-governed mode. CFD simulations confirmed that the length of the SIS assembly was adequate to eliminate the nonuniformity of flow patterns of polymer solutions at the assembly outlet, confirming the uniform characteristics of hollow fibers. This invention is significant in the Indian context since a low-cost affordable solution can now be realized for renal failure patients. With this development, it is important to have an adequate postprocessing strategy to treat these spun membranes in order to enhance their properties. This is explored in the next chapter.

References

1. Bryjak, M., Gancarz, I., Krajciewicz, A., and Piglowski, J. 1996. Air plasma treatment of polyacrylonitrile porous membrane. *Die Angew. Makromol. Chem.* 234: 21–29.
2. Koda, Y. 2011. Clinical outcomes of the high-performance membrane dialyzer. *Contrib. Nephrol.* 173: 58–69.
3. Shakaib, M., Ahmed, I., Yunus, R.M., and Idris, A. 2013. Morphology and thermal characteristics of polyamide/monosodium glutamate membranes. *Int. J. Polym. Mater. Polym. Biomater.* 62: 345–350.
4. Yatzidis, H., Garidi, M., Vassilikos, C., Mayopoulou, D., and Akilas, A. 1964. An improved method for the simple and accurate colorimetric determination of urea with Ehrlich's reagent. *J. Clin. Pathol.* 17: 163–165.
5. Depner, T.A. 2005. Hemodialysis adequacy: Basic essentials and practical points for the nephrologist in training. *Hemodialysis Int.* 9: 241–254.
6. Yamashita, A.C. 2011. Mass transfer mechanisms in high-performance membrane dialyzers. *Contrib. Nephrol.* 173: 95–102.
7. Ambalavanan, S., Rabetoy, G., and Cheung, A.K. 1999. High efficiency and high flux hemodialysis. *Atlas of Diseases of the Kidney*, 5: 1-10, Wiley-Blackwell, NJ.
8. National Institute of Diabetes and Digestive and Kidney Diseases. n.d. Treatment methods for kidney failure: Hemodialysis. Retrieved June 15, 2015, from http://www.niddk.nih.gov/health-information/health-topics/kidney-disease/hemodialysis/Pages/facts.aspx.
9. Pontoriero, G., Pozzoni, P., Andrulli, S., and Locatelli, F. 2003. The quality of dialysis water. *Nephrol. Dial. Transplant.* 18: 21–25.
10. Thakur, B.K. and De, S. 2012. A novel method for spinning hollow fiber membrane and its application for treatment of turbid water. *Sep. Purif. Technol.* 93: 67–74.
11. Roy, A., Vincent, L., and De, S. 2016. Indigenous ultra low cost technology for spinning haemodialysis grade hollow fibre membranes. *BMJ Innov.* 2(2): 84–92.
12. Roy, A., Vincent, L., Rao, S., and De, S. 2016. Low cost spinning and fabrication of high efficiency (he) haemodialysis fibers and method thereof. US Patent 20160008528.
13. Rafizah, W.A.W. and Ismail, A.F. 2008. Effect of carbon molecular sieve sizing with poly (vinyl pyrrolidone) K-15 on carbon molecular sieve–polysulfone mixed matrix membrane. *J. Membr. Sci.* 307(1): 53–61.
14. Venkatasubbu, G.D., Ramasamy, S., Avadhani, G.S., Ramakrishnan, V., and Kumar, J. 2013. Surface modification and paclitaxel drug delivery of folic acid modified polyethylene glycol functionalized hydroxyapatite nanoparticles. *Powder Technol.* 235: 437–442.
15. Wang, C., Feng, L., Yang, H., Xin, G., Li, W., Zheng, J., Tian, W., and Li, X. 2012. Graphene oxide stabilized polyethylene glycol for heat storage. *Phys. Chem. Chem. Phys.* 14(38): 13233–13238.
16. Zeng, M., Fang, Z., and Xu, C. 2004. Effect of compatibility on the structure of the microporous membrane prepared by selective dissolution of chitosan/synthetic polymer blend membrane. *J. Membr. Sci.* 230(1): 175–181.
17. Gutul, T., Rusu, E., Condur, N., Ursaki, V., Goncearenco, E., and Vlazan, P. 2014. Preparation of poly(N-vinylpyrrolidone)-stabilized ZnO colloid nanoparticles. *Beilstein J. Nanotechnol.* 5(1): 402–406.

18. Lowenthal, M.S., Khanna, R.K., and Moore, M.H. 2002. Infrared spectrum of solid isocyanic acid (HNCO): Vibrational assignments and integrated band intensities. *Spectrochim. Acta A* 58(1): 73–78.
19. Mushtaq, A., Mukhtar, H.B., and Shariff, A.M. 2014. FTIR study of enhanced polymeric blend membrane with amines. *Res. J. Appl. Sci. Eng. Technol.* 7(9): 1811–1820.
20. Filimon, A., Avram, E., and Stoica, I. 2014. Rheological and morphological characteristics of multicomponent polysulfone/poly (vinyl alcohol) systems. *Polym. Int.* 63(10): 1856–1868.
21. Ronco, C., Ghezzi, P.M., and Bowry, S.K. 2004. Membranes for hemodialysis. In *Replacement of Renal Function by Dialysis*: 301–323. Springer, Dordrecht, the Netherlands.

Postprocessing of Dialysis Membranes

The essence of strategy is choosing what not to do.

—**Michael E. Porter**

6.1 Introduction

The previous chapters helped in establishing the suitable polymeric composition for hemodialysis membranes and in developing an extrusion mechanism for spinning of clinical grade hemodialysis hollow fibers. The present chapter explores the potential of posttreatment strategies for spun fibers to improve their properties. Annealing is a common practice in this regard.

In 2011, 211 million dialyzers were sold around the world and almost 90% of these used synthetic membranes. In Japan, dialyzers are divided into five types (Classes I–V) on the basis of beta-2 microglobulin (B2M) clearance, with classes IV and V being high-performance dialyzers (HPMs) with high B2M clearance. Ninety percent of dialysis treatment is carried out using these HPMs. Therefore, synthetic HPMs are the focal point of current research efforts. As discussed previously, various complications can arise due to dialysis using high-efficiency (HE) membranes. Dialysis-related amyloidosis (DRA) is a common disease associated with retention of middle-molecular-weight solutes, like B2M (14,000 Da), which the HE membranes cannot clear from blood due to small pore sizes. HPMs have adequate pore size to clear the blood of these middle molecules, and hence the Japanese Society of Dialysis Therapy has observed that patients treated with HPMs enjoyed a better life and had reduced dialysis sessions.[1]

Another class of dialysis grade membranes is high cutoff membranes (HCOs), which have 20,000–55,000 Da molecular weight cutoff (MWCO) and are used to treat patients suffering from sepsis by removal of interleukins and cytokines. HCOs are also used to treat patients suffering from acute pancreatitis and multiple myeloma.[2] However, retaining essential proteins such as human serum albumin (molecular weight 67,000 Da) sets an upper limit to membrane pore size.

Hence, it is important to synthesize HPMs and HCOs and understand the effect of annealing on the membranes. Consequently, in order to investigate the effect of annealing on these membranes, the physical and biological responses have to be evaluated. Thus, a fresh look at membranes and their surface properties focuses on the following points: (1) annealing and its effect on the pore sizes of membranes, (2) change in surface properties due to annealing, (3) changes in cytocompatibility and hemocompatibility, (4) identifying fundamental parameters on which these two depend, (5) whether they can be tuned to yield desired cytocompatibility and hemocompatibility of membranes.

6.2 Experimental Section

6.2.1 Solution Preparation and Membrane Synthesis

The polymeric solution was prepared as per the method described in Chapter 4. The composition for HPMs was 18 wt% polysulfone (PSf), 3 wt% polyvinylpyrrolidone (PVP), and 3 wt% polyethylene glycol (PEG) dissolved in dimethyl formamide (DMF). For HCOs, the ratio of PSf:PVP:PEG was 18:3:0. Hollow-fiber membranes of the same composition were spun using the technology described in the previous chapter. Flat-sheet membranes were cast as described in earlier chapters.

6.2.2 Membrane Annealing

The synthesized membranes were taken out of the gelation bath and subjected to wet and dry heating. Oven heating was carried out at 150°C, and samples of membranes were taken out at 1, 3, and 5 minutes. Beyond 5 minutes, the membranes started to char and hence were not considered. Conversely, wet heating of membranes was carried out at 90°C in a water bath and the samples were taken out at 30-, 45-, and 60-minute intervals.

6.2.3 Membrane Characterization

Cast and spun membranes were characterized in terms of permeability, MWCO, pore diameter, mechanical strength, porosity, and contact angle. The procedure was as described in the previous chapter.

6.2.4 Surface Polarizability and Energy of Adhesion

In order to understand this section, a brief introduction to the science of cell adhesion is required. The basic fundamentals of the liquid droplet and the sessile drop method[3] to measure the contact angle on a surface have been discussed previously. The same principle for measuring the contact angle of various solvents on polymeric membranes has been followed in this section. The rationale behind the exercise is discussed as follows.

Cell adhesion is the primary step of cell interaction with a substrate. This precedes subsequent events of cell spreading, cell proliferation, and eventually cell differentiation. Hence, to explore cell proliferation on polymer membranes, as discussed in Chapter 2, it is imperative to explore cell adhesion first. The fundamental property that dictates cell adhesion on polymeric substrates is its polarizability. Polarizability of a substrate is a complicated manifestation of interfacial free energies between the liquid and the solid surface as well as the surface free energy of the solid surface alone.

In order to find out how polarizable a surface is, various polar liquid droplets and their contact angles on a given surface are measured.[4] The variation in interaction energies of the liquid–solid interface induces polarity on the surface of the substrate. This phenomenon has ramifications in various other observations like cell adhesion. Various surface interactions of a liquid and the substrate are presented as[4]

$$\frac{\gamma_L}{\sqrt{\gamma_L^D}}\left(1+\cos\theta\right) = 2\sqrt{\gamma_S^D} + k_s\frac{\gamma_L^P}{\sqrt{\gamma_L^D}} \tag{6.1}$$

where

γ_L is the liquid surface tension
γ_L^D is the dispersion component of the surface tension
γ_S^D is the dispersion component of the polymer substrate
γ_L^P is the polar component of the surface tension
k_s is surface polarizability

A straight line between $\dfrac{\gamma_L}{\sqrt{\gamma_L^D}}\left(1+\cos\theta\right)$ and $\dfrac{\gamma_L^P}{\sqrt{\gamma_L^D}}$ produces an intercept equal to $2\sqrt{\gamma_S^D}$ and a slope of k_s. k_s is related to the cell adhesion capabilities of a substrate. The typical values of γ_L, γ_L^D, and γ_L^P are reported in Table 6.1 for the set of polar solvents chosen for the study, that is, water, glycerol, formamide, and ethylene glycol.[4]

The energy of adhesion can be explained with the help of interfacial energies. If two materials, each with different surface free energies, are brought in contact with each other, then the energy of adhesion (W_{ad}) measures the strength of contact between the two. The energy of adhesion of blood with the polymer membranes was measured with help of the blood contact angle (following the sessile drop method) using a goniometer, which is based on a

Table 6.1 Liquid Surface Tension and Its Components for Various Polar Solvents

	Water	Glycerol	Formamide	Ethylene Glycol
γ_L (mJ/m²)	72.8	63.4	58.2	48
γ_L^D (mJ/m²)	21.8	37	39.5	29
γ_L^P (mJ/m²)	51	26.4	18.7	19

method developed by Wu.[5] The energy of adhesion can be related to the hemo-compatibility of a substrate.

6.2.5 Nanoindentation

Nanoindentation of the membrane samples was carried out using a nanoin-denter (TI 950 Triboindenter, Hysitron Inc., USA) by determining the load versus displacement curve. At first, the instrument was air-calibrated with a constant electrostatic force. A total number of 10 points, spaced 5 μm apart, were employed on the surface of flat-sheet membranes to understand the plastic work done by the indentor. A maximum load of 100 μN was applied with an unloading and loading time of 10 seconds. Nanoindentation can be correlated with the extent of cell adhesion on a surface.

Again, to understand the methodology of this characterization, it is important to understand the principle of operation of the machine. Nanoindentation is a powerful tool to evaluate mechanical properties (like the ones discussed in Chapter 4 for ultimate tensile strength) at the microscale or nanoscale. The various parameters of materials studied with this instrument are elastic modulus, hardness, viscous parameters, and so on. The advantage of this technology lies in the ability to analyze very small material volumes. The goal in the majority of studies is to extract hardness and elastic modulus of a material from load–displacement measurements. A specified load is applied via a nanoindenter; typical ones are depicted in Figure 6.1. Diamond, which is very hard (and brittle), is a common choice of material for a nanoindenter.

Each type of nanoindenter has its own set of advantages and disadvantages. Spherical indenters are enjoying increased popularity, since they provide smooth transition from elastic to elastic–plastic contact. In a typical nanoindentation test, force and depth of penetration load recorded as load are varied cyclically from zero to maximum and back to zero. If deformation is plastic, a residual impression of force is left on the surface of the specimen, whereas for elastic deformation there is no residual impression and the material recovers it original shape. Figure 6.2a depicts the basic functioning

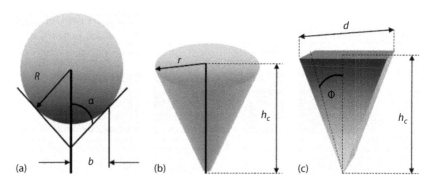

Figure 6.1   Few types of nanoindentors: (a) spherical, (b) conical, and (c) vickers.

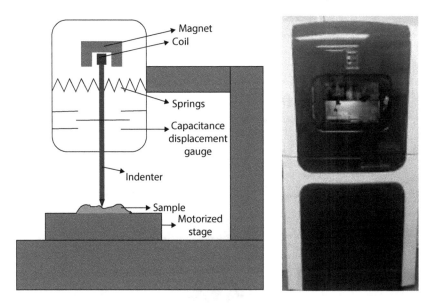

Figure 6.2   (a) Components of a nanoindentor and (b) typical nanoindentor.

of such a machine. The instrument consists of a motorized stage, on which the sample is mounted, and the stage can be moved precisely. Load coils are used to impose load, and capacitive gauges are used to measure displacement. Figure 6.2b depicts a typical nanoindentation machine.

6.2.6 Atomic Force Microscope

Synthesized flat-sheet membranes were cut into smaller sizes (~0.5 cm^2) and were placed on stubs with the help of double-sided adhesion tapes. The surface properties of these membranes were then investigated using an atomic force microscope (MultiMode 8 AFM system Brukker, Sweden) in a tapping mode at room temperature.

It is important to understand the basic working principle of an AFM. Figure 6.3a and b depict the operation of a typical AFM. An AFM has three basic elements: a probe, a positioner, and a processing unit. The probe (or tip) is attached to a rectangular cantilever. The way the probe works is depicted in Figure 6.3b. Once the tip is in the vicinity of the surface, there is a deflection in the cantilever due to interatomic interaction between the tip and the surface. This deflection that results in the force between the tip and the surface can be measured. An AFM can operate in three modes, viz., contact, non-contact, and tapping. In the contact mode, the tip is in actual contact with the sample and the surface of the sample actually pulls or pushes the cantilever. This deflection is fed to a DC amplifier, which in turn applies a voltage to keep the tip in contact with the surface. As a result, a map of the surface with respect to a known sample is created by varying the voltage and the position of the tip. The noncontact mode is used in cases of handling fragile samples.

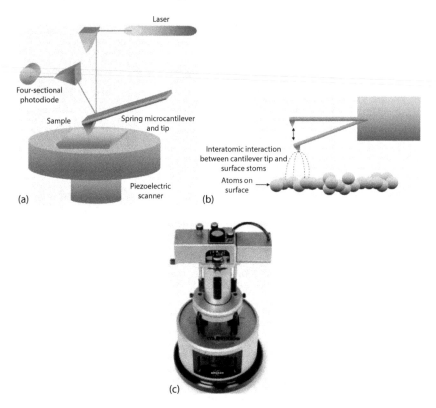

Figure 6.3 (a) Typical working principle of AFM, (b) schematic tip of interaction with surface, and (c) typical AFM.

In this mode, the tip does not touch, but hovers over the sample, measuring van der Waal's forces between the surface and the tip. The tip oscillates, and an AC detector measures change in amplitude, phase, and frequency of oscillations. These forces are strongest within the first few nanometers from the surface. In the third mode, that is, the tapping mode, a piezoelectric crystal causes oscillations at 50–500 Hz with an amplitude of 20–200 nm. The tip scans the surface, but instead of dragging effect, the tip just taps the surface. Here, phase images (also known as contrast images) are obtained and they can be combined with topographical images to reveal surface features along with adhesive properties, friction interactions, and so on. A typical AFM is depicted in Figure 6.3c.

6.2.7 3D Surface Profilometry

Hollow-fiber membranes were subjected to surface profilometry, using a 3D surface profilometer (Bruker, ContourGT-I). Commercial grade (Fresenius) along with laboratory-spun dialysis membranes of 3 mm length were cut, and the surface roughness was measured.

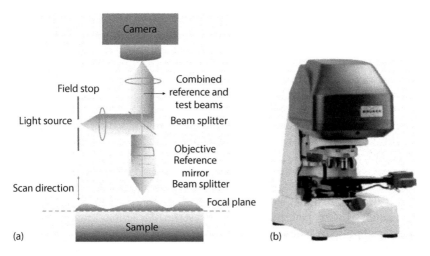

Figure 6.4 (a) Working principle of a profilometer and (b) typical surface profilometer.

Surface profilometers use the principle of interference of light to measure the surface roughness of a sample. The wave properties of light are used to compare the optical path difference between a reference surface and a sample. This is depicted in Figure 6.4a. A light beam is split into two parts, and one part is reflected from the reference mirror itself to the camera. The second part is reflected from the surface of the sample, which is passed through the focal plane of the objective lens of the microscope. The two beams are recombined and interference patterns are observed, that is, alternate dark (destructive interference) and light (constructive interference) bands are examined. From this map of the surface, and the data obtained from a reference surface (e.g., a mirror), the height differences are obtained and a 3D map is generated with the help of a camera. This can be used as an effective tool to measure the surface finish, roughness, and shape of many surfaces, which can be paper, metallic, plastic, and others. A typical surface profilometer is depicted in Figure 6.4b.

6.2.8 Cytocompatibility and Hemocompatibility

Cytocompatibility and hemocompatibility tests (metabolic activity cell proliferation (MTT), cell adhesion, platelet adhesion, hemolysis, thrombus formation, and protein adsorption) were carried out as reported in Chapter 2. The only other test that was performed in addition to these was complement activation. Complement activation during a dialysis session can lead to severe allergic reactions, causing a patient to have anaphylactic shocks, which is extremely dangerous.

6.2.8.1 Complement Activation

Blood collected from a healthy individual was heparinized and immediately centrifuged for 15 minutes at 1500 relative centrifugal force to separate plasma.

Human plasma was placed in the lumen of a 10 cm long piece of prepared membranes and five pieces of 10 cm long commercial Fresenius fibers having equal surface area. It was allowed to stand for 30 minutes at 37°C. Diluents from a commercial MicroVue SC5b-9 Plus EIA kit were used and the recovered plasma was diluted 10 times. The fluid phase concentration of the terminal complement complex (TCC, SC5b-9) was measured using commercially available enzyme-linked immunosorbent assay (ELISA) kit (Quidel Corporation, San Diego, USA). The procedure was followed as per manufacturer's instructions.

It is important to discuss and understand complement activation in more detail. Complement is a group of 30 plasma proteins acting in cohesion that are responsible for the human body's defense mechanism, for example, tolerating infection.[6,7] Craddock et al.[8] first reported complement activation in 1977. It was also reported that severe leukopenia occurred in all patients undergoing hemodialysis.[9] Complement activation is a series of reactions brought about by a group of proteins leading to inflammatory responses of the human body. The proteins are named from C1 through C9. Of these, C3 is particularly responsible for inflammatory response as well as phagocytosis. The defense mechanism of the body is triggered only after a specific enzyme called "C3 convertase" is formed. There are two proposed pathways through which the activation of the complex occurs. These are depicted in Figure 6.5a and b, representing the classical and alternative pathways, respectively. The classical pathway consists of protein C1 attaching to the bacterial cell wall, thereby forming an antigen–antibody complex. This complex cleaves C2 into C2a and C2b and C4 into C4a and C4b. C2b and C4b combine to form a protease called C3b convertase. This in turn cleaves C3 to C3a and C3b. In the alternative pathway, C3b binds to factor B, factor D, and properdin forming C3 convertase, which then carries out a function similar to that in the classical pathway of cleaving C3 to C3a and C3b. It can be generalized that whichever pathway is followed for complement activation, the formation of an antigen–antibody complex is vital for triggering complement activation. Kazatchkai and Carreno[10] discussed the cleavage of an internal thioester bond of C3 molecule leading free oxygen atoms to undergo a transesterification reaction with an –OH or –NH$_2$ group. This covalent bond causes C3 to attach to the foreign surface, thereby triggering a cascading reaction. The individual roles of C3a and C3b as well as C5 are described in Figure 6.5c and d. A comprehensive review article on the subject is available by Ratnoff, and readers can refer to it for more biological details.[11]

6.3 Results and Discussion

6.3.1 Membrane Synthesis, Annealing, and Surface Roughness

The properties of the membranes obtained after synthesis as well as after annealing are depicted in Tables 6.2 and 6.3, respectively. The HPM has an MWCO of 16 kDa, and the HCO has an MWCO of 44 kDa. The HPM has a nominal pore size of 3.67 nm, and the HCO has a nominal pore size of 6.45 nm. The HPM has lower permeability than the HCO, which is expected due to lower pore sizes and porosities than in the case of HCO.

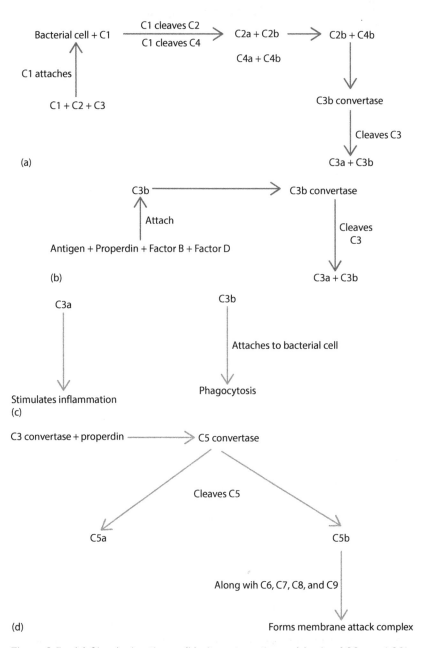

Figure 6.5 (a) Classical pathway, (b) alternate pathway, (c) role of C3a and C3b, (d) role of C5.

Table 6.2	Membrane Properties after Synthesis		
Sl. No.	Membrane Properties	HPM	HCO
1.	Hydraulic permeability ($\times 10^{10}$, m/s Pa)	3.0	6.0
2.	MWCO (kDa)	16	44
3.	Nominal pore size (nm)	3.67	6.45
4.	Porosity (%)	65	70

The properties of annealed membranes are depicted in Table 6.3. It is observed that wet heating increases the pore sizes of the membranes, whereas dry heating decreases them. Such an observation has been noted in another study, details of which were, however, not explored.[12] It is also observed that changes in the physical properties of membranes saturate after 45 minutes of wet heating. Dry heating has a more pronounced effect on the change in membrane properties with reduction in pore size of the membranes from 6.45 nm to 5.01 nm (for HCO) and 3.67 nm to 2.82 nm (for HPM). On the basis of this analysis, it can be concluded that the optimum time for treatment for dry heating would be 3 minutes and for wet heating 45 minutes.

All the membranes are referred to in the subsequent sections by the following nomenclature: HPM45W (HPM 45 minutes wet heating), HPM3D (HPM 3 minutes dry heating), HCO45W (HCO 45 minutes wet heating), and HCO3D (HCO 3 minutes dry heating). The untreated HPM is denoted by 16N and HCO by 44N. The AFM images of the membranes are presented in Figure 6.6. It is observed that the wet and dry heated membranes have relatively less rough surface than the untreated membranes. Thus, HPM45W and HCO45W and HPM3D and HCO3D are smoother than 16N and 44N, respectively.

6.3.2 Contact Angle, Energy of Adhesion, and Surface Polarizability

The contact angles of various solvents obtained on the membrane surface are depicted in Table 6.4. It is observed that annealed membranes have improved surface wettability characteristics than the untreated membranes. The dry heated membranes have the lowest contact angle and hence the best surface wettability (for both HPM and HCO).

Figure 6.7 depicts the straight line between $\dfrac{\gamma_L}{\sqrt{\gamma_L^D}}\left(1+\cos\theta\right)$ and $\dfrac{\gamma_L^P}{\sqrt{\gamma_L^D}}$. The various surface polarizability (k_s) values and the dispersion component of polymer substrate (γ_S^D) values are reported in Table 6.5.

It is evident that k_s values for annealed membranes are higher than for native ones. Two more important points can be noted here. First, it is observed that the k_s value for HPM3D is highest among the six membranes. Second, wet heated membranes have lower k_s values than the dry heated ones. The trend is similar for γ_S^D values as well. The effects of these observations are discussed in subsequent sections.

Table 6.3 Membrane Properties of Annealed Membranes

SI. No.	Membrane Properties	HPM						HCO					
		Wet Heating			Dry Heating			Wet Heating			Dry Heating		
	Time of annealing (minutes)	30	45	60	1	3	5	30	45	60	1	3	5
1.	Hydraulic permeability ($\times 10^{10}$) m/s Pa	3.5	4.0	4.0	3.0	2.0	2.0	6.7	7.0	7.0	5.5	5.0	5.0
2.	MWCO (kDa)	16	20	20	16	10	10	44	55	55	35	28	28
3.	Nominal pore size (nm)	3.67	4.16	4.16	3.67	2.82	2.82	6.45	7.30	7.30	5.68	5.01	5.01
4.	Porosity (%)	65	66	66	65	65	65	70	71	72	70	69	69

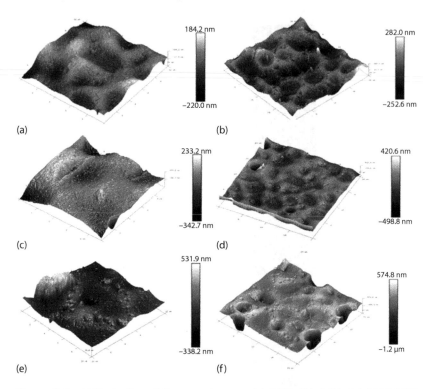

Figure 6.6 AFM studies of the membranes: (a) 16N, (b) 44N, (c) HPM45W, (d) HCO45W, (e) HPM3D, and (f) HCO3D.

Table 6.4 Contact Angle of the Six Membranes in Various Solvents				
Solvents' Membranes	Water	Glycerol	Formamide	Ethylene Glycol
16N	64.2 ± 0.1	61.2 ± 0.2	52.3 ± 0.1	48.2 ± 0.3
HPM45W	62.3 ± 0.2	58.5 ± 0.3	49.2 ± 0.1	46.1 ± 0.2
HPM3D	60.2 ± 0.2	55 ± 0.2	50 ± 0.2	42 ± 0.3
44N	81.5 ± 0.1	74.3 ± 0.3	65 ± 0.3	63 ± 0.2
HCO45W	79.8 ± 0.1	73.1 ± 0.2	64.3 ± 0.4	60.2 ± 0.3
HCO3D	75 ± 0.3	65 ± 0.2	60 ± 0.2	55 ± 0.3

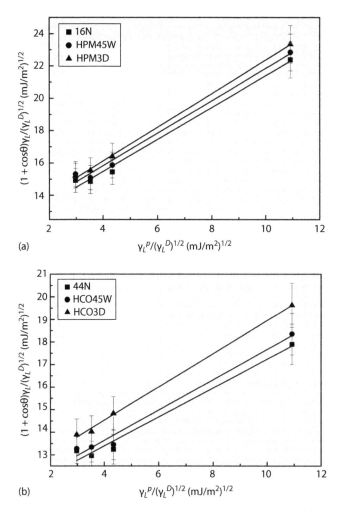

(a)

(b)

Figure 6.7   Surface polarizability diagrams of (a) HPM membranes and (b) HCO membranes.

Table 6.5 k_s and γ_S^D for the Six Membranes		
Membranes	k_s	γ_S^D (mJ/m²)
16N	0.9858 ± 0.01	33.26 ± 0.4
HPM45W	1.0016 ± 0.02	34.97 ± 0.4
HPM3D	1.039 ± 0.01	35.76 ± 0.4
44N	0.6395 ± 0.01	29.37 ± 0.4
HCO45W	0.6708 ± 0.03	30.00 ± 0.4
HCO3D	0.7355 ± 0.02	33.58 ± 0.4

Table 6.6 Energy of Adhesion for Various Membranes with respect to Blood	
Membranes	Energy of Adhesion (mJ/m^2)
16N	87.63 ± 0.5
HPM45W	86.94 ± 0.5
HPM3D	96.24 ± 0.5
44N	93.05 ± 0.5
HCO45W	83.4 ± 0.5
HCO3D	95.61 ± 0.5

The energy of adhesion associated with various membranes and blood has been calculated as presented in the preceding section, and the values are reported in Table 6.6.

This has been calculated with the help of contact angles as discussed earlier.[5,13] It is observed that HPM3D has the highest and HCO45W has the lowest energy of adhesion. Generally, wet heated membranes have lower energy of adhesion than dry heated membranes. These results have implications for hemocompatibility results of membranes and are discussed later.

6.3.3 Nanoindentation Studies

The nanoindentation results are illustrated in Figure 6.8. It is evident that for the same load, native HPM, that is, 16N, is less rigid than HPM45W and HPM3D. Similar is the case for the native HCO (44N) membrane, which shows less rigidity than HCO45W and HCO3D. This fact can be deduced from the hysteresis loss of the surface once the load is removed. The 16N membrane has a hysteresis loss in displacement of 275 nm, whereas for the 44N membrane it is 290 nm. HPM45W has a hysteresis loss of 245 nm and HCO45W has a loss of around the same value of 250 nm. However, HPM3D has a hysteresis loss of 180 nm and HCO3D has a loss of 220 nm. These results affect the extent of cell adhesion and are discussed later.

6.3.4 MTT Assay and Fluorescent and SEM Imaging

The results of MTT assay are depicted in Figure 6.9, and cell adhesion and its scanning electron microscope (SEM) and fluorescent images are illustrated in Figure 6.10. HCO3D exhibits the highest cell proliferation among the HCO membranes, but the highest proliferation, among all membranes, is exhibited by HPM3D. The results for days 1, 2, and 3 follow the same trend. It is generally observed that annealed membranes perform better than the untreated ones. HPM membranes perform better than HCO membranes. There is an increase of 150% in proliferation from day 1 to day 3; however, this jumps to 200%–300% between days 3 and 7. The SEM and fluorescent results exhibit the same trend, as most cells are seen to adhere to the HPM3D membrane and the

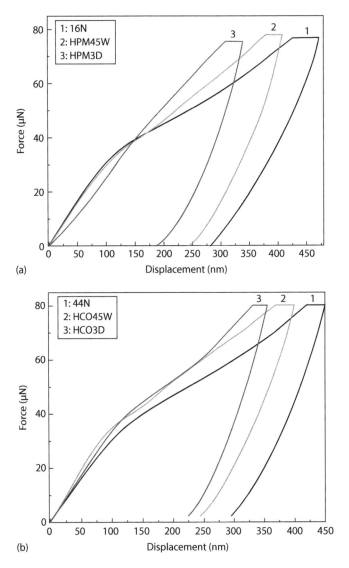

Figure 6.8 Nanoindentation studies of the membranes: (a) HPM membranes and (b) HCO membranes.

least to the 44N membrane. Dry heated membranes exhibit more cells adhering to the surface than the wet heated ones, with the overall observation being that more number of cells adhere to annealed membranes than to native ones.

6.3.5 Hemolysis Assay

The hemolysis results of the membranes are depicted in Figure 6.11. Considering positive control as reference it is observed the HCO membranes perform better than HPM. With respect to positive control, all the membranes perform better.

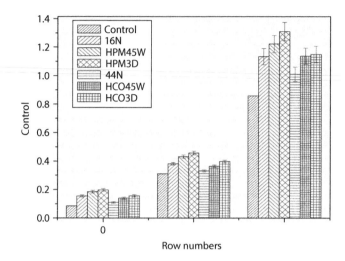

Figure 6.9   MTT assay for the membranes.

6.3.6 Protein Adsorption

Protein adsorption results are presented in Figure 6.12. It is observed that all the membranes exhibit much less protein adsorption levels than the commercial membranes used as control. It is important to note that 44N has the highest protein adsorption and HPM3D has the lowest level of adsorption. Generally, dry heated membranes experience lower protein adsorption than wet heated ones.

6.3.7 Platelet Adhesion

Platelet adhesion on the membranes is illustrated in Figure 6.13. It is evident that HPM membranes have lower platelet adhesion than the HCO category. A closer look at the images reveals that the dry heated membranes have cracks on their surface that are absent in wet heated membranes. Wet heated membranes have lower platelet adhesion than dry heated ones; HPM45W is significantly better than untreated 44N. HCO3D is better performing with respect to HPM3D.

6.3.8 Thrombus Formation

From Figure 6.14, it is clear that the degree of thrombus formation is the highest for HPM3D and the least for HCO45W. HCO3D performance is close to that of HPM3D. HPM45W and HCO45W are closer to the untreated 16N and 44N membranes, respectively. This is consistent with the observation for the energy of adhesion and is discussed later.

6.3.9 Complement Activation

An analysis of Figure 6.15 reveals the complement activation for the membranes. It is evident that HPM3D has the highest levels of complement activation

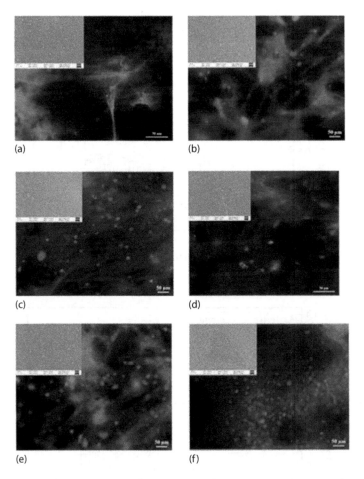

Figure 6.10   Fluorescent and SEM images of cells on membrane surface (day 7 results): (a) 16N, (b) HPM45W, (c) HPM3D, (d) 44N, (e) HCO45W, and (f) HCO3D.

with respect to the other five membranes. With respect to control, HCO45W is the best-performing membrane. In general, wet heated membranes have lower levels of complement activation than dry heated ones. These observations are linked to the energy of adhesion results and are discussed later.

6.3.10 Overall Summary of the Results and Its Implication

The first aspect to discuss about this study is the physical properties of the membranes. Polymers like PVP and PEG help a membrane engineer tailor the pore sizes. Membranes are formed by the process of phase inversion. In this study, non-solvent-induced phase inversion[14–16] has been employed as has been discussed in the previous chapters. In this regard, it is important to note that although both PVP and PEG are hydrophilic polymers, yet they work

153

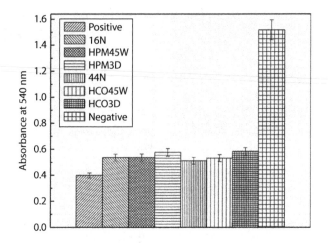

Figure 6.11   Hemolysis assay of the membranes.

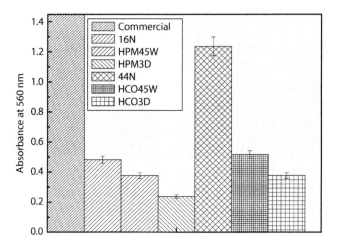

Figure 6.12   Protein adsorption for the membranes.

in an antagonistic manner.[17] PVP induces the formation of pores, whereas PEG helps in constricting them. This interplay manifests in the MWCO and hydraulic permeability values of HCO and HPM membranes (Tables 6.2 and 6.3). HCO membranes have only PVP and this induces quicker demixing of the solution, yielding more open membranes. HPMs have PEG as well in their composition; hence, the viscosity of the solution is more, leading to delayed demixing and hence denser membranes. Thus, the MWCO of HCO is more than that of HPM and, consequently, their porosities and hydraulic permeabilities too follow the same trend.

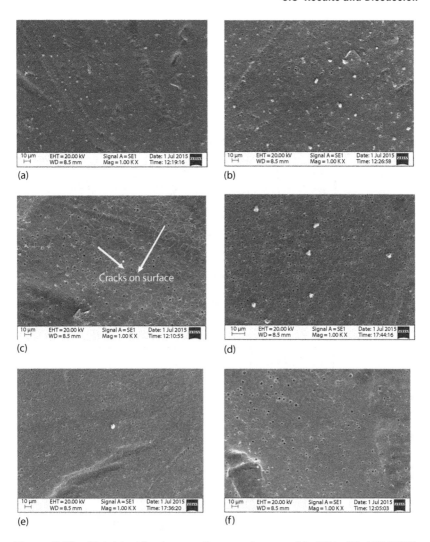

Figure 6.13 Platelet adhesion on the membranes: (a) 16N, (b) HPM45W, (c) HPM3D, (d) 44N, (e) HCO45W, and (f) HCO3D.

Engineers have resorted to annealing as a posttreatment methodology for membranes. However, a detailed study and investigations of such methodologies are not adequately reported. The temperature for the wet heating process was 90°C and for dry heating was 150°C as per previous reports.[18,12] However, this can be explained on the basis of basic annealing principles. Annealing is carried out to remove stress from the material and offer a homogenous structure. The melting point of PSf is 180°C–190°C, and wet heating at 90°C helps the polymeric chains in the membrane matrix attain a minimum Gibbs

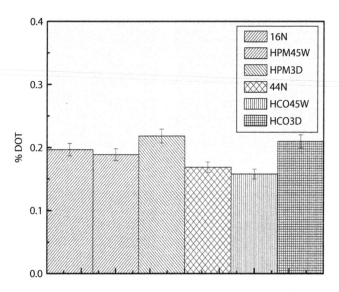

Figure 6.14   Thrombus formation of the membranes.

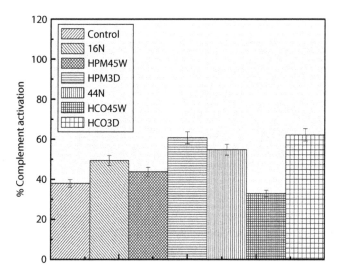

Figure 6.15   Complement activation of the membranes.

free energy, stable configuration. This causes a decrease in surface rough-ness as observed in AFM images (Figure 6.6). For dry heating, there is a rapid exchange of solvent and drying up of membranes, since PSf is primarily a hydrophobic polymer. Hence, it causes shrinkage of polymeric chains, leading to cracks on the surfaces (as seen from SEM images, Figure 6.13).

Surface polarizability (Figure 6.7) has been utilized in this study to understand the nature of a polymeric membrane substrate and its impact on cytocompatibility results. The energy of adhesion has been calculated to relate its application on observed hemocompatibility results. A look at Table 6.5 suggests that the dispersive energy for HCO membranes (29–33 mJ/m^2) is lower than for HPM membranes (around 33–35 mJ/m^2). It automatically follows that the surface polarizability of HPM membranes is higher than that of HCO membranes, which implies that in contact with a liquid, HPM gets more polarized than HCO. Carre (2007)[4] reported that higher the values of k_s, higher would be the cell adhesion capabilities of the substrate for a particular cell line. This corroborates with the findings in this study. The fluorescent and SEM images reveal that membranes with higher k_s values experience higher cell adhesion and vice versa. The 16N membrane has a k_s value of 0.9858, HPM45W has a k_s value of 1.0016, and HCO3D has a k_s value of 1.039; hence, cell adhesion on these three membranes increases in the respective order as observed from Figure 6.6. Similar observations were carried out for 44N, HCO45W, and HCO3D membranes, which have k_s values of 0.6395, 0.6708, and 0.73, respectively, thereby increasing cell adhesion in that order. It can be concluded from these observations that annealed membranes undergo higher cell adhesion than untreated membranes; maximum adhesion is highest in case of dry heated membranes. However, cell adhesion properties are not dependent on only k_s values. The other crucial aspect for cell adhesion on a substrate is surface rigidity.

As depicted in Figure 6.16, cells have the capability to sense various forces including shear stress, compression, and even elasticity.[19] There is a specific mechanism for cells that self-generate forces according to the rigidity of substrates at the focal adhesion complexes (FACs) or the cell adhering sites. FACs are the active sites of cells, controlling adherence to the external environment via mechanosensitivity and mechanotransduction.[19] Once the FACs sense the intensity of force on a substrate, a complicated activation of multiple proteins is initiated. This regulates tension of the actomyosin fibers, resulting in final adhesion of cells. Cells can sense and respond to forces as low as 5 pN. Various cell lines generate a variety of mechanical stimuli for adhesion and proliferation. Fibroblasts and endothelial cells have been reported to adhere and proliferate on a substrate having elastic modulus higher than 3 kPa.[19] Another study revealed that on an anisotropic rigid surface, the cells grow and migrate toward higher rigidity.[19] Hence, regions of high traction forces support better cell proliferation. It can thus be safely concluded that cell adhesion and proliferation are dependent on a complex interplay between the physical characteristics of surface rigidity and the polarizability of the surface.

Keeping the preceding discussion in mind, it will be interesting to revisit the results of fluorescent imaging, SEM (Figure 6.10), and MTT (Figure 6.9) assays. As has been established, the most rigid surface undergoes the highest cell adhesion and vice versa. Hence, it automatically follows that HPM3D and HCO3D (highest k_s in HPM and HCO categories) exhibit the highest proliferation in both the categories. Figure 6.8 indicates that HPM3D is the most

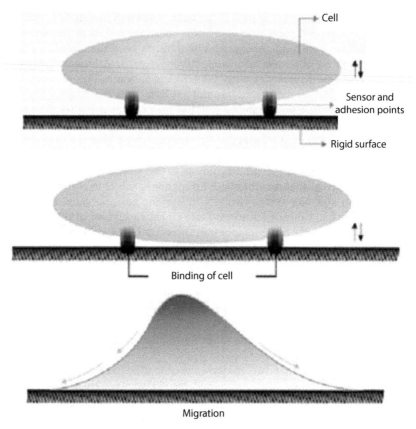

Figure 6.16   Cell adhesion to rigid surfaces.

rigid membrane (and has the highest k_s among all the membranes) and experiences the highest cell adhesion as well as cell proliferation. Consequently, the least rigid membrane 44N has the lowest adhesion and proliferation of cells. Again, a general observation would yield that annealed membranes undergo better cytocompatibility results than the untreated ones, which is supported by the visual depiction of Figure 6.10. Therefore, tuning of these surface properties can lead to hollow-fiber membranes with better biological response.

Similarly, hemocompatibility results can be analyzed based on the energy of adhesion. More energy is required to separate a liquid from a substrate and vice versa, and it is detrimental to dialysis applications because it can lead to blood coagulation/adsorption. It is clear that HCO45W has the least energy of adhesion (implying least energy required to separate blood from its surface) and hence a lower tendency of blood adsorption. Conversely, HPM3D has the highest energy of adhesion (implying highest energy required to separate blood from surface) and hence is prone to blood coagulation the most. This is a significant result as far as the ensuing discussion on blood compatibility is concerned. All the hemocompatibility results, that is, hemolysis, platelet adhesion, thrombus formation, and complement activation, are related to the interaction of surface with blood. Only protein adsorption is dependent on a more significant parameter and that is hydrophilicity (Figure 6.12). It is easily observable that HPM3D is consistently a poor performer among all the synthesized membranes. HPM3D experiences the highest degree of hemolysis, platelet adhesion, thrombus formation, as well as complement activation (Figures 6.11 and 6.13 through 6.15). On the other end of the spectrum, HPM45W is the best-performing membrane with the lowest levels of hemolysis, platelet adhesion, and complement activation. The general observation in this discussion is that, with wet heat treatment, membranes perform better than the native ones as well as dry heated ones.

This is a very significant observation in case of dialysis membranes or for that matter any surfaces dealing with blood interactions. For hemodialysis, hemocompatibility issues are more pertinent than cytocompatibility. A closer look at Figure 6.13 reveals that cracks develop on the surface of dry heated membranes. The combined effect of these two observations leads to the common conclusion that dry heated membranes are not suitable for dialysis applications.

In order to verify the rationale of annealing developed in this study, lab-developed HPM and HCO fibers are compared with commercial Fresenius fibers using optimized postannealing technology (heating at 90°C for 45 minutes). It is evident from Figure 6.17 that spun fibers when subjected to the developed annealing process yield a smoother surface (Ra = 7.9 µm and 10.1 µm) as compared to commercially available F6 membranes (Ra = 13.1 µm).

6.4 Conclusion

This chapter provides a closure to the technology of hemodialysis membrane synthesis. It started with the synthesis of a suitable material, then moved on to developing an extrusion mechanism to spin clinical grade hollow-fiber membranes. Finally, suitable postprocessing of the spun fibers that have been developed was discussed in this chapter. Fundamental insights into cytocompatibility and hemocompatibility of dialysis membrane surfaces have also been developed. It was observed that k_s and surface rigidity are related to cell proliferation, whereas energy of adhesion is related to hemocompatibility. It was also found out that dry heating enhances the cytocompatibility of polymeric membranes as compared to wet heated membranes. This trend is reversed when the hemocompatibility results are explored. It was seen that wet heated membranes had lower energy of adhesion as compared to dry heated ones. This induces lower extent of complement activation and platelet adhesion on wet heated membranes than their dry heated counterparts. Moreover, due to the rapid rates of solvent transfer during dry heating, the

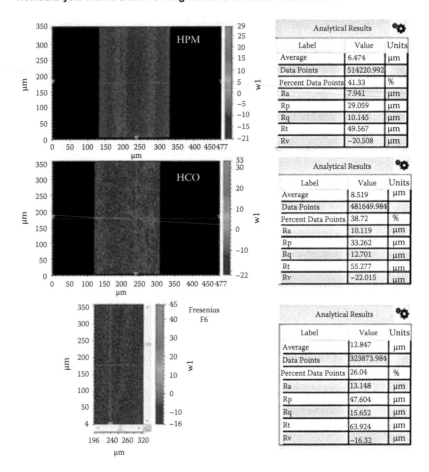

Figure 6.17   3D surface profilometry of Fresenius F6, HPM45W, and HCO45W hollow-fiber membranes.

surface of membranes develops cracks, which is highly undesirable for applications like hemodialysis. Finally, from the point of engineering, it is realized that wet heating is a better option and easier to retrofit to the spinning process discussed in the previous chapter. Hence, drawing a holistic picture, it is easy to conclude that wet heating (45 minutes at 90°C) is a better option as a posttreatment strategy for dialysis membranes. This chapter is also very important when it comes to the holistic development of a technology. The heating of membranes needs energy and this shoots up the operating costs. The implications of this have an effect on the cost of production, and this has been explored in the last chapter on commercialization. In fact, value addition in a product always leads to an increase in product cost, but it is acceptable for sensitive applications like hemodialysis.

References

1. Vanholder, R.C. 2010. Advantages of new hemodialysis membranes and equipment. *Nephron Clin. Pract.* 114: 165–172.
2. Villa, G., Zaragoza, J.J., Sharma, A., Neri, M., De Gaudio, A.R., and Ronco, C. 2014. Cytokine removal with high cut-off membrane: Review of literature. *Blood Purif.* 38: 167–173.
3. Hamid, N.A.A., Ismail, A.F., Matsuura, T., Zularism, A.W., Lau, W.J., Yuliwati, E., and Abdullah, M.S. 2011. Morphological and separation performance study of polysulfone/titanium dioxide (PSF/TiO₂) ultrafiltration membranes for humic acid removal. *Desalination* 273: 85–92.
4. Carre, A. 2007. Polar interactions at liquid/polymer interfaces. *J. Adhes. Sci. Technol.* 21: 961–981.
5. Wu, S. 1971. Calculation of interfacial tension in polymer systems. *J. Polym. Sci. C* 34: 19–30.
6. Kinoshita, T. 1991. Overview of complement biology. *Immunol. Today* 12: 291–294.
7. Jahns, G., Haeffner-Cavallion, N., Nydegger, U.E., and Kazatchkine, M.D. 1993. Complement activation and cytokine production as consequences of immunological bioincompatibility of extracorporeal circuits. *Clin. Mater.* 14: 303–336.
8. Craddock, P.R., Fehr, J., Dalmasso, A.P., Brigham, K.L., and Jacob, H.S. 1977. Haemodialysis leukopenia. *J. Clin. Invest.* 59: 879–888.
9. Leonard, S., Kaplow, M.D., John, A., and Goffinet, M.D. 1968. Profound neutropenia during the early phase of hemodialysis. *JAMA* 203: 1135–1137.
10. Kazatchkai, M.D. and Carreno, M.P. 1988. Activation of the complement system at the interface between blood and artificial surfaces. *Biomaterials* 9: 30–35.
11. Ratnoff, W.D. 1980. A war with the molecules: Louis Pillemer and the history of properdin. *Perspect. Biol. Med.* 23: 638–657.
12. Barzin, J., Feng, C., Khulbe, K.C., Matsuura, T., Madaeni, S.S., and Mirzadeh, H. 2004. Characterization of polyethersulfone hemodialysis membrane by ultrafiltration an atomic force microscopy. *J. Membr. Sci.* 237: 77–85.
13. Owens, D.K. and Wendt, R.C. 1969. Estimation of the surface free energy of polymers. *J. Appl. Polym. Sci.* 13: 1741–1747.
14. Arockiasamy, D.L., Alam, J., and Alhoshan, M. 2013. Carbon nanotubes-blended poly(phenylene sulfone) membranes for ultrafiltration applications. *Appl. Water Sci.* 3: 93–103.
15. Liao, Y., Farrell, T.P., Guillen, G.R., Li, M., Temple, J.A., Li, X.G., Hoek, E.M., and Kaner, R.B. 2014. Highly dispersible polypyrrole nanospheres for advanced nanocomposite ultrafiltration membranes. *Mater. Horiz.* 1(1): 58–64.
16. Liao, Y., Wang, X., Qian, W., Li, Y., Li, X., and Yu, D.G. 2012. Bulk synthesis, optimization, and characterization of highly dispersible polypyrrole nanoparticles toward protein separation using nanocomposite membranes. *J. Colloid Interface Sci.* 386(1): 148–157.
17. Panda, S. and De, S. 2014. Preparation, characterization and performance of ZnCl₂ incorporated polysulfone (PSF)/polyethylene glycol (PEG) blend low pressure nanofiltration membranes. *Desalination* 347: 52–65.
18. Bowry, S.K., Emanuele, G., and Vienken, J. 2011. Contribution of polysulfone membranes to the success of convective dialysis therapies. *Contrib. Nephrol. Basel Karger.* 173: 110–118.
19. Hervy, M. 2010. Modulation of cell structure and function in response to substrate stiffness and external forces. *J. Adhes. Sci. Technol.* 24: 963–973.

7

Mathematical Modeling
of Dialysis

As far as the laws of mathematics refer to reality, they are not certain; and as far as they are certain, they do not refer to reality.

—Albert Einstein

7.1 Introduction

Any technological development has various points to be explored and understood, and in the process it presents scope for improvement. Understanding and developing the membrane is a complicated process and involves making a cartridge and fitting in an extracorporeal circuit to perform hemodialysis. In this regard, each of the mentioned steps deserves volumes of materials to be written in order to aid comprehension. However, mathematical modeling is one approach that can help the reader understand the basic principles and realize the interdependence of parameters that are part of the process. A lot of literature is available on mathematical modeling of hemodialysis, and the authors have classified and summarized the various approaches as illustrated in Figure 7.1.

7.2 Mathematical Models

7.2.1 Intrinsic to Membrane

7.2.1.1 Membrane Pore Modeling

The pores of membranes play a very crucial and significant role in transporting solutes across them. Solute transport through pores is represented by the pore diffusion model, the basic assumptions of which are

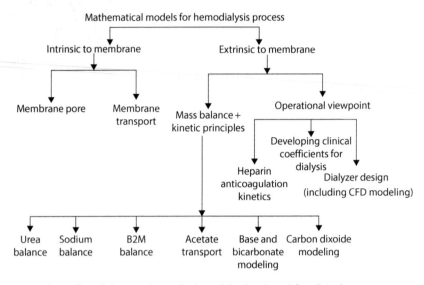

Figure 7.1   Classifying mathematical models developed for dialysis.

1. Solutes cannot permeate through dry membranes; thus, transport occurs only through wet membranes.

2. The rate of transport depends on solute distribution within the solution to the membrane surface.

3. The force of friction (impedance to transport) is dependent on the pore configuration.

This has been applied to early modeling of peritoneal dialysis.[1,2] The restriction factor for diffusion of solute is expressed as[3]

$$\frac{a_{eff}}{a_0} = \frac{\left(1-\alpha\right)^{9/2}}{1-0.395\alpha+1.0616\alpha^2} \tag{7.1}$$

where
$\alpha = r_s/r_p$, r_s is the Stoke's radius of the solute, and r_p is the pore radius
a_{eff} is the effective area of the pore cross section
$a_0 = \pi r_p^2$ is the nominal surface area of pore

7.2.1.2 Membrane Transport Modeling

It is imperative to discuss the fundamental Staverman–Kedem–Katchalsky–Spiegler model regarding solvent (J_v) and solute (J_s) transport through a flat-sheet permselective membrane, which is expressed as[4-7]

$$J_v = L_P\left(\Delta P - \sigma RT\Delta C\right) \tag{7.2}$$

$$J_s = P_D \Delta C + (1 - \sigma) J_v C_M \tag{7.3}$$

where

  C is the concentration of the solute
  P_D is the diffusive permeability (defined in Chapter 5)
  σ is the Staverman reflection coefficient
  C_M is the mean transmembrane concentration of solute
  ΔP and ΔC are transmembrane pressure and concentration differences, respectively
  L_P is the hydraulic permeability of the membrane
  R is the universal gas constant
  T is the absolute temperature
  C_M can be evaluated according to the following equation with boundary conditions C_1 and C_2:

$$C_M = (1 - f) C_1 + f C_2 \tag{7.4}$$

$$f = \frac{1}{\lambda J_v} - \frac{1}{\exp(\lambda J_v) - 1} \tag{7.5}$$

$$\lambda = \frac{1 - \sigma}{P_D} \tag{7.6}$$

where $Pe = \lambda |J_v|$ is the Peclet number, representing the ratio of convective to diffusive transport.

As an exercise, the detailed derivation of Equations 7.2 and 7.3 is outlined in Appendix 7A.2.

Deen[1] proposed a variant of Equation 7.6 to express the solute flux:

$$J_s = \varepsilon \left[v K_C C - D_\infty K_D \frac{\partial c}{\partial r} \right] \tag{7.7}$$

where

  ε is the porosity
  C is local solute concentration
  v is local fluid velocity
  D_∞ is the Brownian diffusion coefficient for the solute
  K_C and K_D are hindrance factors for convection and diffusion, respectively

They capture the interaction of solutes with pore walls. Morti et al.[8] considered a typical asymmetric hollow-fiber membrane structure and integrated Equation 7.7 through each layer of dense and porous zones resulting in the expression of J_s as

$$J_s = \frac{\varepsilon v \varphi_1 K_{C1} \left[C_L \exp\left(\dfrac{K_{C1} v \delta_1}{K_{D1} D_\infty} \right) - C_S \exp\left(-\dfrac{K_{C2} v \delta_2}{K_{D2} D_\infty} \right) \right]}{\left(\dfrac{\varphi_1 K_{C1}}{\varphi_2 K_{C2}} \right) \exp\left(\dfrac{K_{C1} v \delta_1}{K_{D1} D_\infty} \right) - \left(\dfrac{\varphi_1 K_{C1}}{\varphi_2 K_{C2}} \right) \exp\left(-\dfrac{K_{C2} v \delta_2}{K_{D2} D_\infty} \right) - 1} \tag{7.8}$$

where

"1" and "2" denote dense and porous layers, respectively

K_C and K_D are hindrance factors for convective and diffusive transports, respectively

ε is local membrane porosity

v is local fluid velocity

φ is equilibrium partition coefficient

δ_1 and δ_2 are the thickness of the dense and porous layers, respectively

D_∞ is the Brownian diffusion coefficient

C_L and C_S are solute concentrations of blood and dialysate, respectively, and are related to bulk concentrations by

$$\varphi = \frac{C_{r=0}}{C_L} = \frac{C_{r=\delta_m}}{C_S} \tag{7.9}$$

where δ_m is the membrane thickness.

Steady state is assumed for deriving Equation 7.8, where solute flux through each region (1 and 2) is the same and is equal to solute flux through the membrane.

In another model, Legallais et al.[9] used an expression to evaluate local transmembrane solute flux J_s, which was similar to Equation 7.8 and based on the following assumptions:

1. Hemodiafilter operates under steady-state condition.

2. There is no cross-sectional variation of flow rates, concentrations, and pressure, and these vary only along the length of the filter.

3. Axial diffusion is negligible with respect to axial convection.

4. The concentration at the membrane surface and the surrounding bulk phase are equal.

Hence, the following equation for flux was derived:

$$J_s = K_0 \left(C_B(x) - C_D(x) \right) + J_v(x) S_\infty \left(f_B(x) C_B(x) - f_D(x) C_D(x) \right) \tag{7.10}$$

where

$$f_B = \frac{\exp \left[\left(\frac{S_\infty J_v}{P_D} \right) + \frac{J_v}{K_B} + \frac{J_v}{K_D} \right]}{\left(1 - S_\infty \right) \exp \frac{J_v}{K_D} \left[\exp \left(\frac{S_\infty J_v}{P_D} \right) - 1 \right] + S_\infty \exp \left[\frac{S_\infty J_v}{P_D} + \frac{J_v}{K_B} + \frac{J_v}{K_D} \right] - S_\infty - \frac{K_0}{J_v S_\infty}} \tag{7.11}$$

$$f_D = \frac{-1}{\left(1 - S_\infty \right) \exp \frac{J_v}{K_D} \left[\exp \left(\frac{S_\infty J_v}{P_D} \right) - 1 \right] + S_\infty \exp \left[\frac{S_\infty J_v}{P_D} + \frac{J_v}{K_B} + \frac{J_v}{K_D} \right] - S_\infty + \frac{K_0}{J_v S_\infty}} D \tag{7.12}$$

where

K_0 is overall mass transfer coefficient under zero flux conditions
S_∞ is the sieving coefficient at infinite flux
C_B is the solute concentration in blood
C_D is the solute concentration in the dialysate
P_D is solute diffusive permeability through the membrane
K_B is the mass transport coefficient in blood
K_D is the mass transport coefficient in the dialysate

Raff et al.[10] used a variant of this model by including the effect of concentration polarization, which was neglected in the model. The wall concentrations were expressed as a function of volumetric flux and bulk concentrations:

$$C_w(x) = C_B(x)\exp\left(\frac{J_v(x)}{k(x)}\right) \qquad (7.13)$$

where

C_w is the concentration at the membrane wall
C_B is the bulk concentration of solute
k is the mass transfer coefficient on the blood side

An advanced model considering both convection and diffusion along the axial direction was proposed by Moussy et al.[11] If a careful investigation is made, the chronological order of model development for hemodialysis can easily be arrived at. In fact, the requirement for better models arose once researchers understood hemodialysis better. Preliminary models developed way back in the 1970s and 1980s, like in the work of Jaffrin et al.,[12] Kunimoto et al.,[13] and Sigdell et al.,[14] discussed just the accumulation of larger solutes on the membrane surface. Better understanding of hemodialysis made scientists realize that the accumulation of larger solutes does not just lead to the formation of a loose layer over the membrane, but also leads to adsorption of solutes. This was explored by Gachon et al.[15] As a result, the detrimental effects of adsorption on the flux and hydraulic permeability were studied by Belfort et al.,[16] Langsdorf and Zydney,[17] and Meireles et al.[18] At the same time, adsorption leading to pore blockage of membranes was modeled by Nilsson and Hallstrom.[19] Later on, advanced two-layer theory on the formation of a secondary gel layer was examined by Boyd and Zydney[20] and Langsdorf and Zydney.[17]

7.2.2 Extrinsic to Membrane

The discussion in this section will be a combined effort where we will be revisiting clinical parameters that were introduced in Chapter 2 in a more practical manner. The true relevance of modeling of hemodialysis systems and definitions of clinical parameters like Kt/V, and K_oA will be clear to the readers in this section.

7.2.2.1 Revisiting Clearance

In a nutshell, a look at Figure 7.2 will help in recalling the basic physics of hemodialysis. It has been established in the previous chapters that state-of-the-art hemodialysis includes a combined convective diffusive mechanism employed to remove uremic toxins from blood resulting from a pressure differential applied across the membranes, and thereby producing a net ultrafiltration flux. Hence, it is clear that if water loss occurs from blood to the dialysate side, then the inlet and outlet blood flow rates will not be equal. This simple logic leads to the following formulation:

$$Q_{UF} = Q_{Bi} - Q_{B0} \tag{7.14}$$

where
Q_{UF} is the ultrafiltration rate
Q_{Bi} is the inlet blood flow rate
Q_{B0} is the outlet blood flow rate; all the units are in mL/min

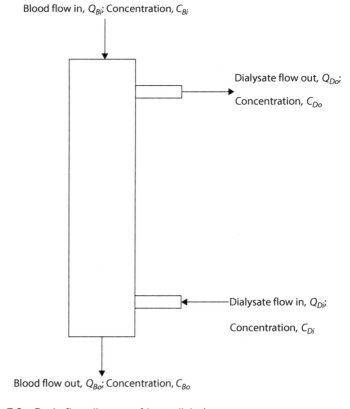

Figure 7.2   Basic flow diagram of hemodialysis.

With the definition of clearance (Section 2.3.6, Kt/V definition, where K is clearance), the following equation results[21]:

$$K = Q_{Bi}\underbrace{\left(\frac{C_{Bi} - C_{B0}}{C_{Bi}}\right)}_{1} + Q_F\underbrace{\left(\frac{C_{B0}}{C_{Bi}}\right)}_{2} \qquad (7.15)$$

where

C_{Bi} is inlet blood concentration
C_{B0} is outlet blood concentration

Term 1 of Equation 7.15 is the diffusive part (denoted by K_B) and Term 2 is the convective part. The subscript B is for the blood side. Similar equations exist for the dialysate side as well[21]:

$$K = Q_{Di}\underbrace{\left(\frac{C_{Di} - C_{D0}}{C_{Di}}\right)}_{1} + Q_F\underbrace{\left(\frac{C_{D0}}{C_{Di}}\right)}_{2} \qquad (7.16)$$

As in Equation 7.15, Terms 1 and 2 denote the diffusive and convective parts, respectively. The diffusive part can be denoted as K_D. Thus, dialysate clearance can also be written as[22]

$$K_D = Q_{Di}\left(\frac{C_{Di} - C_{D0}}{C_{Di}}\right) \qquad (7.17)$$

PROBLEM 7.1
Given that during the hemodialysis procedure of a patient, the arterial urea concentration is 110 mg/dL, venous urea concentration is 30 mg/dL, with a typical blood flow rate of 300 mL/min. If the patient has to lose 2 kg during a 4-hour run, then calculate clearance of urea.

Solution: The arterial concentration indicates $C_{Bi} = 110$ mg/dL and venous blood indicates $C_{B0} = 30$ mg/dL. The 2.5 kg is the water weight that the patient has to lose. Hence, the ultrafiltration rate is (weight gain × 1000)/dialysis time = (2 × 1000)/240 = 8.33 mL/min. The clearance as per Equation 7.15 is

$$K = Q_{Bi}\left(\frac{C_{Bi} - C_{B0}}{C_{Bi}}\right) + Q_{UF}\left(\frac{C_{B0}}{C_{Bi}}\right) = 300\left(\frac{110 - 30}{110}\right) + 8.33\left(\frac{30}{110}\right) = 220 \text{ mL/min.}$$

7.2.2.2 Revisiting K_{UF}

Clinically, K_{UF} of a hemodialyzer denotes the volume (mL) of fluid that is transferred to the dialysate side per hour when 1 mmHg of transmembrane pressure (TMP) is applied.[23,24] The average water flux (Q_{UF}) is defined as

$$Q_{UF} = \frac{\iint (\Delta P - \Delta \pi) K_{UF} \, dA}{A} \qquad (7.18)$$

where $\Delta\pi$ denotes the oncotic pressure.[25,26] If K_{UF} is constant over surface area A and TMP is constant, then[21]

$$Q_{UF} = K_{UF} \int_L (\Delta P - \Delta\pi) \frac{dx}{L} = K_{UF} \times \text{TMP} \qquad (7.19)$$

PROBLEM 7.2

Calculate the ultrafiltration rate required for 3 kg of fluid during a 4-hour dialysis period and the TMP if a hemodialyzer with $K_{UF} = 5\ \dfrac{\text{mL}}{\text{h. mmHg}}$.

Solution: The ultrafiltration rate $Q_{UF} = (3 \times 1000)/4 = 750\ \text{mL/h}$, considering density of the fluid to be 1000 kg/m³.

So, the TMP required will be $Q_{UF}/K_{UF} = 750/5 = 150\ \text{mmHg}$.

7.2.2.3 Revisiting k_0A

k_0A is derived from first principles of mass transfer as discussed later. However, to understand this, it is important to look at Figure 7.3. It represents the linear decrease in concentration on the blood side and an increase in the dialysate side, in a countercurrent flow. Using the concept of countercurrent heat transfer, the mass transfer in hemodialysis can be expressed in terms of overall mass transfer coefficient (k_0) and membrane area (A), and the rate of solute transfer to the dialysate side ($\dot{m}$) is expressed as

$$\dot{m} = k_0 A \left[\frac{\Delta C_i - \Delta C_0}{\ln\left(\dfrac{\Delta C_i}{\Delta C_0}\right)} \right] \qquad (7.20)$$

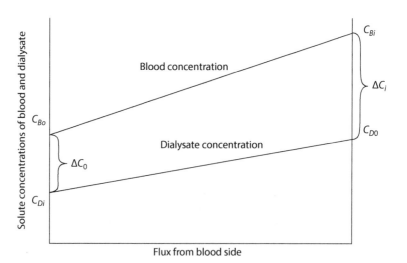

Figure 7.3 Blood and dialysate concentration profiles as function of flux.

where

ΔC_i is the inlet concentration difference between blood and dialysate

ΔC_0 is the outlet concentration difference between blood and dialysate sides

Overall mass transfer coefficient k_0 is expressed as

$$\frac{1}{k_0} = \frac{1}{k_f} + \frac{1}{k_D} + \frac{\delta_m}{D_m}$$ (7.21)

where

k_f and k_D are mass transfer coefficients in the feed and dialysate, respectively

δ_m is the membrane thickness

D_m is solute diffusivity in the membrane phase

In fact, for hemodialyzers, the clearance K can also be related to the overall mass transfer coefficient (k_0) as[27]

$$K = Q_B \left[\frac{\exp\left[\dfrac{k_0 A}{Q_B}\left(1 - \dfrac{Q_B}{Q_D}\right)\right] - 1}{\exp\left[\dfrac{k_0 A}{Q_B}\left(1 - \dfrac{Q_B}{Q_D}\right)\right] - \dfrac{Q_B}{Q_D}} \right]$$ (7.22)

PROBLEM 7.3

Given that for $Q_B = 300$ mL/min and $Q_D = 600$ mL/min, a dialyzer has an in vitro clearance of 200 mL/min. Find out the in vitro clearance if Q_D is increased to 800 mL/min. What is the percentage increase in clearance? Consider the Daugirdas and Depner adjustment[27]:

Dialysate Flow Rate	Multiplication Factor
500	1
600	1.033
700	1.066
800	1.099
900	1.132
1000	1.165

Solution: The $k_0 A$ for $Q_B = 300$ mL/min and $Q_D = 600$ mL/min is

$$k_0 A = \frac{300 \times 600}{300 - 600} \ln\left[\frac{1 - \dfrac{200}{300}}{1 - \dfrac{200}{600}}\right] = 415 \text{ mL/min.}$$

A look at the Daugirdas and Depner table indicates that increasing the Q_D to 800 mL/min would induce a correction factor of 1.099; hence, $k_0A_{adjusted} = 415 * 1.099 = 457.06\, \text{mL/min}$. Hence, the clearance (modified) would be

$$K = Q_B \left[\frac{\exp\left[\frac{k_0A}{Q_B}\left(1-\frac{Q_B}{Q_D}\right)\right]-1}{\exp\left[\frac{k_0A}{Q_B}\left(1-\frac{Q_B}{Q_D}\right)\right]-\frac{Q_B}{Q_D}} \right] = 300\frac{1.14}{1.64} = 208.65\ \text{mL/min}.$$

Thus, increasing the Q_D to 800 mL/min increases the clearance by only 4%.

7.2.2.4 Pharmacokinetics and Revisiting *Kt/V*

There are two types of models discussing urea removal and they are the single-pool and double-pool models. These are basically kinetic models taking into account the classical mass balance approach. This is depicted in Figure 7.4.

The assumption in this model is that total body water is a single and fixed pool with no urea generation. Urea is generated by the liver (at a rate G), which is supplemented by the kidney, which clears it (K_RC; the product of K_R and C yields the removal of urea in time t) continuously and intermittently with a dialyzer (K_DC). The volume distribution of urea inside the human body is V and the concentration of urea in the volume is C. Thus, applying simple mass balance principles:

$$\text{Rate of input} - \text{Rate of output} + \text{Rate of generation}$$
$$- \text{Rate of consumption} = \text{Rate of accumulation} \tag{7.23}$$

Thus,

$$\frac{d(VC)}{dt} = G - KC \tag{7.24}$$

where $K = K_R + K_D$. Considering V to be constant, Equation 7.24 is an ordinary differential equation and can be solved to give the following solution:

$$C = C_0 \exp\left(-\frac{Kt}{V}\right) + \frac{G}{K}\left[1 - \exp\left(\frac{-Kt}{V}\right)\right] \tag{7.25}$$

If it is assumed that urea generation during dialysis is negligible, then Equation 7.24 becomes

$$\frac{d(VC)}{dt} = -KC \tag{7.26}$$

Solving Equation 7.26:

$$C = C_0 \exp\left(-\frac{Kt}{V}\right) \tag{7.27}$$

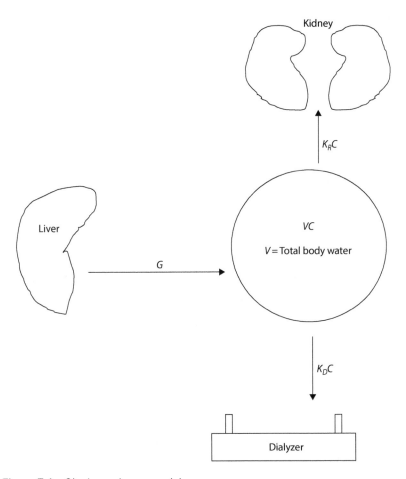

Figure 7.4 Single-pool urea model.

Taking logarithm of both sides of Equation 7.27, we have

$$-\frac{Kt}{V} = \ln\left(\frac{C_0}{C_t}\right)$$ (7.28)

Equation 7.28 is also known as Lowrie's formula.[28]

PROBLEM 7.4
Calculate the postdialysis blood urea nitrogen (BUN) if predialysis BUN is 130 mg/dL, postdialysis urea distribution volume is 40 L, the dialyzer used has a clearance of 220 mL/min, and residual renal clearance is 10 mL/min. Assume urea generation rate at 9 mg/min and dialysis time of 210 minutes.

Solution: Using the single-pool model, the postdialysis BUN can be calculated as

$$C = 130 \exp\left(-\frac{(2.2 + 0.1) \times 210}{40}\right) + \frac{9}{(2.2 + 0.1)}$$

$$\times \left[1 - \exp\left(\frac{-(2.2 + 0.1) \times 210}{40}\right)\right] = 35 \text{ mg/dL}.$$

Another parameter used to understand dialysis adequacy is urea reduction ratio (*URR*). It is defined as

$$\text{URR} = \left(1 - \frac{C_t}{C_0}\right) \times 100 \tag{7.29}$$

C_t/C_0 is known as urea reduction fraction, denoted by R. Combining Equations 7.28 and 7.29, we have

$$\frac{Kt}{V} = -\ln(1 - R) \tag{7.30}$$

It is recommended that for dialysis adequacy, URR ~ 65[29]; hence, $Kt/V \sim 1.2$. This explains the reason behind maintaining Kt/V at 1.2, as discussed in Chapter 2.

A more generalized approach is considering the state-of-the-art ultrafiltration-coupled dialysis session, leading to loss in volume. Thus, the patient's body is considered to be of nonconstant volume $V(t)$ and urea concentration $C(t)$. This is depicted in Figure 7.5.

Thus, during dialysis

$$\frac{dV}{dt} = -Q_F \tag{7.31}$$

and

$$V_t = V_0 - Q_F t \tag{7.32}$$

where
V denotes the postdialysis value
V_0 denotes the predialysis value
Q_F is the rate of fluid removal

Again, a simple mass balance yields:

$$\left(V_0 - Q_F t\right)\frac{dC}{dt} + C\frac{d\left(V_0 - Q_F t\right)}{dt} = G - KC \tag{7.33}$$

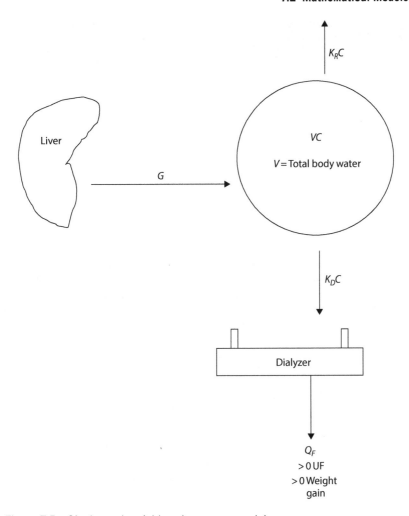

Figure 7.5   Single-pool variable volume urea model.

Solving Equation 7.33, the following expression is obtained[30]:

$$C = C_0 \left(\frac{V_0 - Q_F t}{V_0} \right)^{\frac{K - Q_F}{Q_F}} + \frac{G}{K - Q_F} \left[1 - \left(\frac{V_0 - Q_F t}{V_0} \right)^{\frac{K - Q_F}{Q_F}} \right] \quad (7.34)$$

There are various models available in the literature, which are variants of the single-pool urea model discussed earlier. A few of them are listed in Table 7.1[30]:

An advanced version of the single-pool urea model is the double-pool model. The assumption is that there are two distinct compartments in which

Table 7.1 Mathematical Models for Kt/V Calculation

Model	Expression
Lowrie Model[28]	$\dfrac{Kt}{V} = \ln\left(\dfrac{C_0}{C_t}\right)$
Jindal Model[31]	$\dfrac{Kt}{V} = \left[0.04 \times \left(\dfrac{C_0 - C_t}{C_0}\right) \times 100\right] - 1.2$
Keshavaiah Model[32]	$\dfrac{Kt}{V} = 1.62 \times \ln\left(\dfrac{C_0}{C_t}\right)$
Calzavara Model[33]	$\dfrac{Kt}{V} = \dfrac{2 \times (C_0 - C_t)}{C_0 + C_t}$
Daugirdas First Generation Model[34,35]	$\dfrac{Kt}{V} = -\ln\left(R - 0.008 \times t - f \times \dfrac{UF}{W}\right)$ where K is the clearance (mL/min), t is the time (min), V is the volume of water content (mL), R is the ratio of the postdialysis to predialysis BUN, f is the fudge factor, UF is the ultrafiltration volume per dialysis (mL), W is the postdialysis body weight of the patient (kg)
Daugirdas Second Generation Model[36]	$\dfrac{Kt}{V} = -\ln(R - 0.008 \times t) + (4 - 3.5R)\dfrac{UF}{W}$ where K is the clearance (mL/min), t is the time (min), V is the volume of water content (mL), R is the ratio of the postdialysis to predialysis BUN, f is the fudge factor, UF is the ultrafiltration volume per dialysis (mL), W is the postdialysis body weight of the patient (kg)

the urea is distributed. They are the extracellular and intracellular pools, as illustrated in Figure 7.6.

It is seen that urea is generated by the liver at a time-dependent rate of $G(t)$ to the extracellular fluid (ECF) compartment of volume $V_e(t)$ and concentration $C_e(t)$. The exchange of urea occurs within the extracellular and intracellular volumes as presented in Figure 7.7.

Hence, the governing equations would be

$$\frac{dm_1}{dt} = -K_D C_1 - K_R C_1 + K_{12}(C_2 - C_1) + G \tag{7.35}$$

$$\frac{dm_2}{dt} = -K_{12}(C_2 - C_1) \tag{7.36}$$

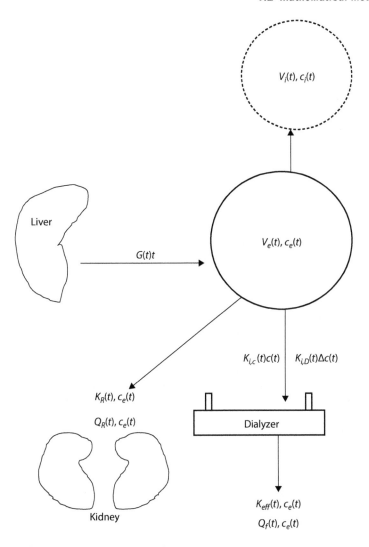

Figure 7.6   Double-pool urea model.

where
Indices 1 and 2 represent the central and peripheral compartments, respectively
m is the solute mass
K_{12} is the intercompartmental clearance
K_R is residual renal clearance
K_D is dialyzer clearance
G is the generation rate of the solute
C is the concentration of the solute

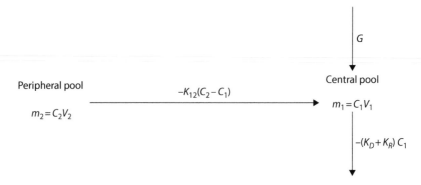

Figure 7.7   Simplified double-pool urea model.

As in the case of urea, there are kinetic models for bicarbonate, such as the body acid–base concentration model, the extracellular water model, the creatinine model, along with the sodium and beta-2 microglobulin (B2M) model.

7.3 Mathematical Method to Find Postdialysis Urea Concentration

To understand the drop in urea concentration in blood, the standard method is to sample blood at 30 and 60 minutes after a dialysis session.[37] However, this may pose a problem and indeed there are various practical issues. The main problem lies in the fact that Kt/V is overestimated by measuring the pre- and postdialysis blood urea concentrations. This phenomenon is also known as postdialysis urea rebound (PDUR). This is due to the intercompartmental equilibration of urea (in line with the double-pool model) occurring 30–90 minutes after standard hemodialysis sessions[38,39] or 60 minutes after 8-hour-long hemodialysis sessions[40]; 12%–40% error in judgment of dialysis adequacy occurs if PDUR is ignored. A solution to this problem, keeping practicality in mind, was suggested by Smye et al.[39]:

$$C_{eq}^{smye} = C_0 \exp\left[\frac{-t_d}{t_s} \ln \frac{C_s}{C_{td}} \right] \tag{7.37}$$

where
C_{eq}^{Smye} is the Smye relation for equilibrated postdialysis BUN
C_0 is the predialysis BUN
C_s is the middialysis BUN (usually after 70 minutes of hemodialysis)
C_t is BUN concentration at the end of hemodialysis
t_d is the duration of hemodialysis
t_s is the sampling time of C_s

Hence, PDUR is expressed as

$$\text{PDUR} = \frac{C_{eq} - C_t}{C_t} \times 100\% \tag{7.38}$$

PROBLEM 7.5

Calculate equilibrated postdialysis urea concentration for the given data:

Predialysis BUN = 140 mg/dL, intradialytic urea concentration (after 90 minutes) = 65 mg/dL, postdialysis urea concentration = 44 mg/dL. The total dialysis duration is 240 minutes. Also calculate the PDUR.

Solution: Using Smye formula:

$$C_{eq}^{Smye} = C_0 \exp\left[\frac{-t_d}{t_s}\ln\frac{C_s}{C_t}\right] = 140\exp\left[\frac{-240}{90}\ln\frac{65}{44}\right] = 49.5 \text{ mg/dL}$$

and

$$PDUR = \frac{C_{eq} - C_t}{C_t}\times 100\% = \frac{49.5 - 44}{44}\times 100\% = 12.5\%$$

Hence, it becomes imperative to understand and estimate the equilibrated dialysis dose $\left(\dfrac{Kt}{V}\right)_{eq}$, keeping in mind PDUR. The standard method to determine this is the Daugirdas et al.[41] formula:

$$\left(\frac{Kt}{V}\right)_{eq} = \frac{\left(\dfrac{Kt}{V}\right)_{sp}}{\ln\left(\dfrac{C_0}{C_t}\right)}\ln\left(\frac{C_0}{C_{eq,True}}\right) \qquad (7.39)$$

where

$\left(\dfrac{Kt}{V}\right)_{sp}$ is the single-pool $\dfrac{Kt}{V}$

$C_{eq,\,True}$ is the true equilibrium concentration (mg/L)

Another method is the Smye method[39]:

$$\left(\frac{Kt}{V}\right)_{eq}^{Smye} = -\ln\left(R_{eq}^{Smye} - 0.008\times t\right) + \left(4 - 3.5\times R_{eq}^{Smye}\right)\times\frac{UF}{W} \qquad (7.40)$$

where

$R_{eq}^{smye} = \dfrac{C_{eq}^{Smye}}{C_0}$

K is the clearance (mL/min)

t is the time (min)

V is the volume of water content (mL)

UF is the ultrafiltration volume per dialysis (mL)

W is the postdialysis body weight of the patient (kg)

Tattersall et al.[42] proposed another formula:

$$\left(\frac{Kt}{V}\right)_{eq}^{Tatt} = \left(\frac{Kt}{V}\right)_{sp}\left[\frac{t}{t+B}\right] \tag{7.41}$$

where B is the Tattersall's time constant (~35 minutes).

Another method was proposed by Maduell et al.[43]:

$$\left(\frac{Kt}{V}\right)_{eq}^{Mad} = \left[0.906\times\left(\frac{Kt}{V}\right)_{sp}\right] - \left[0.26\frac{K}{V}\right] + 0.007 \tag{7.42}$$

where $\frac{K}{V}$ is per hour $\frac{Kt}{V}$, or the dialysis rate.

7.4 Regional Blood Flow Model

This is an improvement over the conventional double-pool model and captures the postdialysis urea rebound. In this model, organs of the human body are divided into two groups: one that has blood flow rate:water volume ratio >0.2/min in one group and the rest in the other group.[30] The first group is known as the "high flow system" group, and the second group is known as the "low flow system" group. The high flow system consists of the brain, heart, lungs, kidneys, and the portal system. The low flow system consists of muscles, bones, skin, and fat. Figure 7.8 depicts the flow diagram of the model. The following set of equations is used to model regional blood flow (RBF)[30]:

$$V_H = \left(V_0 - Q_{UF}.t\right)f_{VH} \tag{7.43}$$

$$V_L = \left(V_0 - Q_{UF}.t\right)f_{VL} \tag{7.44}$$

$$\frac{dC_H}{dt} = \frac{1}{V_H}\left[Q_H\left(C_{art} - C_H\right) + G\right] \tag{7.45}$$

$$\frac{dC_L}{dt} = \frac{1}{V_L}\left[Q_L\left(C_{art} - C_L\right)\right] \tag{7.46}$$

$$f_{cp} = \left[\frac{Q_{sys} + Q_{UF}}{Q_{sys} + K_d}\right] \tag{7.47}$$

$$C_{art} = f_{cp}\frac{C_H\left(Q_{sys}.f_{QH}. + Q_{UF}.f_{VH}\right) + C_L\left(Q_{sys}.f_{QL}. + Q_{UF}.f_{VL}\right)}{Q_{sys} + Q_{UF}} \tag{7.48}$$

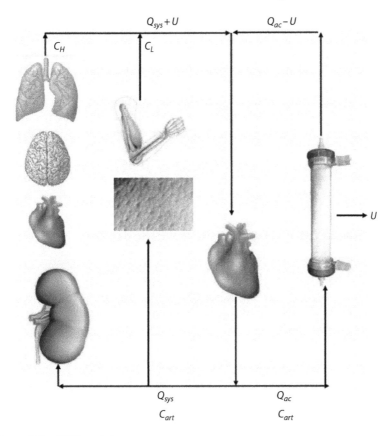

Figure 7.8 RBF model.

where
  V_0 is the initial water volume
  V_H and V_L are volumes of high and low flow systems
  f_{VH} denotes volume fraction of the high flow system
  f_{VL} denotes volume fraction of the low flow system
  f_{QH} denotes flow fraction of the high flow system
  f_{QL} denotes flow fraction of the low flow system
  G is the rate of generation
  Q_{sys} is systemic blood flow
  Q_{UF} is the ultrafiltration rate
  Q_H and Q_L are the blood flow rate in the high and low systems
  t is time
  V_0 is initial water volume
  C_H is concentration in high flow system
  C_L is concentration in low flow system
  C_{art} is concentration in arterial system
  f_{cp} is cardiopulmonary recirculation

181

7.5 Holistic View: Hemodialyzer Modeling

In order to appreciate the marvel of nature, some sobering facts would set the tone for this section. It was seen[44] that even under the high blood flow rates achievable in the present state-of-the-art technologies, mimicking the performance of kidneys was still a dream! This was solely due to the great mismatch between the number of capillaries for filtration in kidneys (100×10^6) as compared to those in dialyzers (100×10^2).[30] In addition to this, the human kidneys have the ability to reabsorb and secrete components (e.g., glucose) very selectively. The membranes of dialyzers have the ability to adsorb certain components but lack the selective capacity to reabsorb.[30] However, the present dialyzers do achieve the target of removal of uremic toxins to an appreciable degree besides maintaining homeostasis.

7.5.1 One-Dimensional Diffusive Model of Hemodialyzer

The efficiency of a hemodialyzer is defined as $\varepsilon = \dfrac{C_{Bi} - C_{Bo}}{C_{Bi} - C_{Di}}$ and $R_1 = \dfrac{Q_B}{Q_D}$, where indices i and o indicate the inlet and the outlet, C denotes concentration of solute and subscripts B and D indicate blood and dialysate sides, respectively, Q_B and Q_D indicate blood and dialysate flow rates, respectively. Hence, the following equation results:

$$\frac{C_{Bi} - C_{Do}}{C_{Bo} - C_{Di}} = \frac{R_1 - \dfrac{1}{\varepsilon}}{1 - \dfrac{1}{\varepsilon}} \tag{7.49}$$

Simplifying Eq. 7.49[30], we have:

$$\varepsilon\left(R_1, \text{NTU}\right) = \frac{1 - \exp\left[-\text{NTU}\left(1 - R_1\right)\right]}{1 - R_1\left[\exp\left(-\text{NTU}\left(1 - R_1\right)\right)\right]} \tag{7.50}$$

where NTU is the number of transfer units defined as $\dfrac{k_0 A}{Q_b}$. The solute transport is thus expressed as

$$Q_b\left(C_{Bi} - C_{Bo}\right) = \varepsilon\left(R_1, \text{NTU}\right) Q_b\left(C_{Bi} - C_{Di}\right) \tag{7.51}$$

7.5.2 Modeling in Cylindrical Coordinates

The limitation of the one-dimensional model is that it is a steady-state analysis without convection and without considering Newtonian flow behavior (flow pattern is depicted in Figure 7.9). Hence, there arose better approaches that led to modeling of hollow-fiber membranes in cylindrical coordinates. For a relevant set of model equations, readers can refer to Azar.[30]

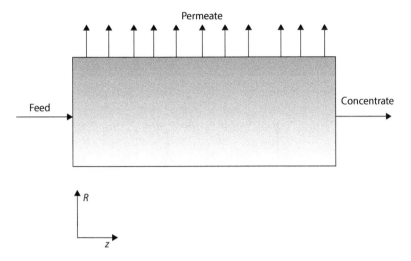

Figure 7.9 Cylindrical coordinate modeling of dialyzers—utilizing axial symmetry.

7.6 Conclusion

Hemodialysis is a well-studied phenomenon when it comes to clinical as well as mathematical approaches. Though it is very difficult to capture the dynamics of the human body in one simple mathematical model, yet researchers have been close enough with their developments thus far. Modeling helps in saving time and in predicting the outcomes of a process like hemodialysis. This in turn proves to be very useful for any dynamic and in-line changes in operating parameters that have to be brought about by the clinician so as to nullify any disturbances creeping into the system, thereby ensuring patient safety and effective hemodialysis treatment.

Appendix 7A

It is to be noted that Equation 7.3 is obtained upon integration of the following phenomenological Equation 7.34.[14]

$$J_s = -p_d \frac{dc}{dx} + (1-\sigma)J_v c \tag{7A.1}$$

Therefore, to understand the correlation between Equations 7.2 and 7.3, the following derivation was carried out. The notations bear the same nomenclature as in the referred publication by Sigdell[14] or

$$J_s - (1-\sigma)J_v c(x) = -p_d \frac{dc}{dx}$$

183

or

$$dx = -p_d \frac{dc}{J_s - (1-\sigma)J_v c(x)} \qquad (7A.2)$$

Let $J_s - (1-\sigma)J_v c(x) = z$; hence, $dc = \dfrac{dz}{(1-\sigma)J_v}$. Thus, (7A.2) reduces to

$$\int_0^\delta dx = -\frac{p_d}{(1-\sigma)J_v} \int_{J_s - (1-\sigma)J_v c_L}^{J_s - (1-\sigma)J_v c_R} \frac{dz}{z}$$

or

$$\delta = -\frac{p_d}{(1-\sigma)J_v} \ln \frac{J_s - (1-\sigma)J_v c_R}{J_s - (1-\sigma)J_v c_L}$$

or

$$\frac{\delta(1-\sigma)J_v}{p_d} = -\ln \frac{J_s - (1-\sigma)J_v c_R}{J_s - (1-\sigma)J_v c_L}$$

or

$$\lambda J_v = -\ln \frac{J_s - (1-\sigma)J_v c_R}{J_s - (1-\sigma)J_v c_L}$$

or

$$\exp(\lambda J_v) = -\frac{J_s - (1-\sigma)J_v c_R}{J_s - (1-\sigma)J_v c_L}$$

or

$$\exp(\lambda J_v) - 1 = \frac{J_s - (1-\sigma)J_v c_R}{J_s - (1-\sigma)J_v c_L} - 1$$

or

$$\frac{1}{\exp(\lambda J_v) - 1} = \frac{J_s - (1-\sigma)J_v c_L}{-(1-\sigma)J_v (c_L - c_R)}$$

or

$$\frac{1}{\lambda J_v} - \frac{1}{\exp(\lambda J_v) - 1} = \frac{1}{\lambda J_v} - \frac{\dfrac{J_s}{P} - \lambda J_v c_L}{\lambda J_v (c_L - c_R)}$$

or

$$f = \frac{\left(c_L - c_R\right) - \dfrac{J_s}{P} - \lambda J_v c_L}{\lambda J_v \left(c_L - c_R\right)}$$

or

$$f \lambda J_v \left(c_L - c_R\right) = \left(c_L - c_R\right) - \frac{J_s}{P} - \lambda J_v c_L$$

or

$$\frac{J_s}{P} = \left(c_L - c_R\right)\left(1 - f \lambda J_v\right) - \lambda J_v c_L$$

or

$$\frac{J_s}{P} = \left(c_L - c_R\right) + \left(\lambda J_v\right)\left(f c_L - f c_R - c_L\right)$$

or

$$\frac{J_s}{P} = \left(c_L - c_R\right) + \left(\lambda J_v\right)\left[c_L\left(1 - f\right) + f c_R\right]$$

or

$$\frac{J_s}{P} = \left(c_L - c_R\right) + \left(\lambda J_v\right) c_M$$

$$J_s = \left(c_L - c_R\right)P + J_v\left(1 - \sigma\right) c_M$$

References

1. Deen, W. 1987. Hindered transport of large molecules in liquid-filled pores. *AIChE J.* 33: 1409–1425.
2. Waniewski, J., Werynski, A., Heimburger, O., and Lindholm, B. 1991. A comparative analysis of mass transport models in peritoneal dialysis. *ASAIO Trans.* 37(2): 65–75.
3. Rippe, B. and Krediet, R.T. 1994. Peritoneal physiology—Transport of solutes. In: *The Textbook of Peritoneal Dialysis*, Gokal, R. and Nolph, K.D. (eds.). Kluwer, Dordrecht, the Netherlands, pp. 69–113.
4. Katchalsky, A. and Curran, P. 1967. *Nonequilibrium Thermodynamics in Biophysics*. Harvard University Press, Cambridge, MA.
5. Spiegler, K.S. and Kedem, O. 1966. Thermodynamics of hyperfiltration (reverse osmosis): Criteria for efficient membranes. *Desalination* 1: 311–326.
6. Curry, F.-R.E. 1984. Mechanics and thermodynamics of transcapillary exchange. In: *Handbook of Cardiology—Cardiovascular System IV*, Renkin, E. M. (ed.). American Physiological Society, Bethesda, MD.

7. Sigdell, J.E. 1974. *A Mathematical Theory for the Capillary Artificial Kidney.* Hippokrates Verlag, Stuttgard, Germany.

8. Morti, S., Shao, J., and Zydney, A.L. 2003. Importance of asymmetric structure in determining mass transport characteristics of hollow fiber hemodialyzers. *J. Membr. Sci.* 224: 39–49.

9. Legallais, C., Catapano, G., Harten, B.V., and Baurmeister, U. 2000. A theoretical model to predict the in vitro performance of hemodiafilters. *J. Membr. Sci.* 168: 3–15.

10. Raff, M., Welsch, M., Göhl, H. et al. 2003. Advanced modeling of highflux hemodialysis. *J. Membr. Sci.* 216: 1–11.

11. Moussy, Y. 2000. Bioartificial kidney. I. Theoretical analysis of convective flow in hollow fiber modules: Application to a bioartificial hemofilter, biotechnology and bioengineering. *Biotechnol. Bioeng.* 68(2): 142–152.

12. Jaffrin, M.Y., Gupta, B.B., and Malbrancq, J.M. 1981. A one-dimensional model of simultaneous hemodialysis and ultrafiltration with highly permeable membranes. *J. Biomech. Eng.* 103: 261–266.

13. Kunitomo, T., Lowrie, E.G., Kumazawa, S., O'Brien, M., Lazarus, J.M., Gottlieb, M.N., and Merrill, J.P. 1977. Controlled ultrafiltration (UF) with hemodialysis (HP): Analysis of coupling between convective and diffusive mass transfer in a new HD-UF system. *ASAIO J.* 23(1): 234–242.

14. Sigdell, J.E. 1982. Calculation of combined diffusive and convective mass transfer. *Int. J. Artif. Organs* 5: 361–372.

15. Gachon, A.M.F., Mallet, J., Tridon, A., and Deteix, P. 1991. Analysis of proteins eluted from hemodialysis membranes. *J. Biomater. Sci. Polym.* 76(2): 263.

16. Belfort, G., Davis, R., and Zydney, A.L. 1994. The behavior of suspensions and macromolecular solutions in crossflow microfiltration. *J. Membr. Sci.* 96: 1–58.

17. Langsdorf, L.J. and Zydney, A.L. 1994. Effect of blood contact on the transport properties of hemodialysis membranes: A two-layer membrane model. *Blood Purif.* 12: 292–307.

18. Meireles, M., Aimar, P., and Sanchez, V. 1991. Effects of protein fouling on the apparent pore size distribution of sieving membranes. *J. Membr. Sci.* 56: 13–28.

19. Nilsson, J. and Hallstrom, B. 1991. Deviations in the fouling resistance of UF membranes due to clean membrane permeability variations. *J. Membr. Sci.* 67: 177–189.

20. Boyd, R.F. and Zydney, A.L. 1998. Analysis of protein fouling during ultrafiltration using a two-layer membrane model. *Biotechnol. Bioeng.* 59: 451–460.

21. Sargent, J.A. and Gotch, F.A. 1996. Principles and biophysics of dialysis. In: *Replacement of Renal Function by Dialysis*, 4th edn., Jacob, C., Kjellstrand, C.M., Koch, K.M., and Winchester, J.F. (eds.). Kluwer Academic Publiher, Dordrecht, the Netherlands, pp. 188–230.

22. Leypoldt, J.K. 2000. Solute fluxes in different treatment modalities. *Nephrol. Dial. Transplant.* 15(Suppl. 1): 3–9.

23. Granger, A., Vantard, G., Vantelon, J., and Perrone, B. 1978. A mathematical approach of simultaneous dialysis and filtration (SDF). In: *Proc. Eur. Soc. Artif. Organs*, vol. 5, pp. 174–177.

24. Cheung, A.K. and Leypoldt, J.K. The hemodialysis membranes: A historical perspective, current state and future prospect. *Semin. Nephrol.* 17(3): 196–213.

25. Ronco, C. 1990. Backfiltration in clinical dialysis: Nature of the phenomenon, mechanisms and possible solutions. *Int. J. Artif. Organs* 13: 11–21.

26. Wolf, A.V., Remp, D.G., KiUey, J.E., and Currie, G.D. 1951. Artificial kidney function: Kinetics of hemodialysis. *J. Clin. Invest.* 30: 1062.

27. Daugirdas, J.T. and Depner, T.A. 1994. A nomogram approach to hemodialysis urea modeling. *Am. J. Kidney Dis.* 23(1): 33–40.

28. Lowrie, E. and Teehan, B. 1983. Principles of prescribing dialysis therapy: Implementing recommendations from the National Cooperative Dialysis Study. *Kidney Int.* 23(Suppl. 13): S113–S122.

29. Health Care Financing Administration (HCFA). 1994. Core Indicators Project initial results: Opportunities to improve care for adult in-center hemodialysis patients. Department of Health and Human Services, Health Care Financing Administration, Health Standards and Quality, Bureau, Baltimore, MD.

30. Azar, A.T. (ed.). 2013. *Modeling and Control of Dialysis Systems.* Vol. 1: Modeling Techniques of Hemodialysis Systems, SCI 404. Springer, Berlin, Germany.

31. Jindal, K.K., Manuel, A., and Goldstein, M.B. 1987. Percent reduction of the blood urea concentration during dialysis (PRU), a simple and accurate method to estimate Kt/V_{urea}. *ASAIO Trans.* 33(3): 286–288.

32. Keshaviah, P.R., Hanson, G.L., Berkseth, R.O., and Collins, A.J. 1988. A simplified approach to monitoring in vivo therapy prescription. *ASAIO J.* 34(3): 620–622.

33. Calzavara, P., Vianello, A., Da Porto, A., Gatti, P.L., Bertolone, G., Caenaro, G., and Dalla Rosa, C. 1988. Comparison between three mathematical models of KT/V. *Int. J. Artif. Organs* 11(2): 107–110.

34. Daugirdas, J.T. 1989. The post: pre-dialysis plasma urea nitrogen ratio to estimate Kt/V and NPCR: Mathematical modeling. *Int. J. Artif. Organs* 12(7): 411–419.

35. Daugirdas, J.T. 1989b. The post: pre-dialysis plasma urea nitrogen ratio to estimate Kt/V and NPCR: Validation. *Int. J. Artif. Organs* 12(7): 420–427.

36. Daugirdas, J.T. 1993. Second generation logarithmic estimates of single-pool variable volume Kt/V: An analysis of error. *J. Am. Soc. Nephrol.* 4(5): 1205–1213.

37. Chirananthavat, T., Tungsanga, K., and Eiam-Ong, S. 2006. Accuracy of using 30-minute post-dialysis BUN to determine equilibrated Kt/V. *J. Med. Assoc. Thai.* 89(Suppl. 2): 54–64.

38. Maduell, F., Sigüenza, F., Caridad, A., López-Menchero, R., Miralles, F., and Serrato, F. 1994. Efecto rebote de la urea: Influencia del volumen de distribución de la urea, tiempo de diálisis y aclaramiento del dializador. *Nefrologia* 14(2): 189–194.

39. Smye, S.W., Dunderdale, E., Brownridge, G., and Will, E. 1994. Estimation of treatment dose in high-efficiency haemodialysis. *Nephron* 67(1): 24–29.

40. Jean, G., Chazot, C., Charra, B., Terrat, J.C., Vanel, T., Calemard, E., and Laurent, G. 1998. Is post-dialysis urea rebound significant with long slow hemodialysis? *Blood Purif.* 16(4): 187–196.

41. Daugirdas, J.T., Greene, T., Depner, T.A., Gotch, F.A., and Star, R.A. 1999. Relationship between apparent (single-pool) and true (double-pool) urea distribution volume. *Kidney Int.* 56(5): 1928–1933.

42. Tattersall, J., Farrington, K., Bowser, M., Aldridge, C., Greenwood, R.N., and Levin, N.W. 1996. Underdialysis caused by reliance on single pool urea kinetic modeling. *J. Am. Soc. Nephrol.* 3: 398.

43. Maduell, F., Garcia-Valdecasas, J., Garcia, H., Hernandez-Jaras, J., Sigüenza, F., Del Pozo, C., Giner, R., Moll, R., and Garrigos, E. 1997. Validation of different methods to calculate KtV considering postdialysis rebound. *Nephrol. Dial. Transplant.* 12(9): 1928–1933.

44. Osada, Y. and Nakagawa, T. 1992. *Membrane Science and Technology.* Marcel Dekker, Inc., New York.

8

Economics of Dialyzer Manufacturing: Commercialization

The Industrial Revolution was another of those extraordinary jumps forward in the story of civilization.

—**Stephen Gardiner**

8.1 Commercialization

As has been discussed previously, dialysis treatment is an expensive process, and, since dialysis cartridges are not manufactured in India, imported cartridges increase the treatment costs further. Each of these dialyzer cartridges costs around Rs. 1500–2000 for the end user, and a kidney failure patient typically requires around 140 such cartridges per year. Non-affordability forces patients to reuse the cartridges, which raises the risks of infection, cardiac arrest, and other complications in subsequent dialysis sessions. Since the present work is perfectly applicable to the Indian context and discusses a potentially marketable product, corresponding financial and business analyses are important.

8.2 Customer Segmentation

The developed technology has tremendous market potential, and likely customer segmentation is presented in Figure 8.1.

Based on this figure, it is clear that the technology can easily be extended to cater to different market segments. It is important to note that

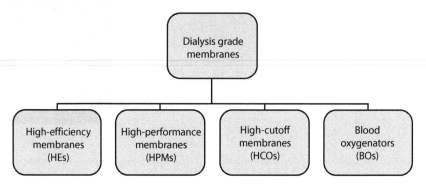

Figure 8.1   Customer segmentation of developed technology.

1. High-efficiency membranes (HEs) are the conventional dialysis membranes.

2. High-performance membranes (HPMs) are the larger pore-sized membranes as discussed in the previous sections.

3. High-cutoff membranes (HCOs) are the membranes used for myeloma patients.

4. Blood oxygenators (BOs) are membranes used for cardiovascular bypass surgery patients.

The technology would prove to be a commercial boon for manufacturers, as the market size is large. While the global market is valued at $75 billion with the growth rate projected at 6% per annum, in India alone there are at least 1.5 lakh renal failure patients, whose number is growing at the rate of 30% per annum. The South-East Asian region, with countries like the Philippines, presents tremendous potential for growth. In fact, the Asia-Pacific region contributes nearly 42% of the total dialysis patients worldwide. Europe, the Middle East, and Africa together contribute 25% of dialysis patients. Moreover, there are over 1.56 million patients suffering from cardiac diseases and 60,000 patients undergo cardiopulmonary bypass. It is reported that around 16,000 myeloma patients are diagnosed every year in India. Hence, from the perspective of business development, this unfortunate scenario presents a huge opportunity for commercialization of the developed end-to-end technology.

8.3 Plant Layout, Design, and Production

With regard to scaling up of the technology, it is imperative to begin the design keeping the most limiting factor, that is, land, in mind. With respect to the available land spaces in a typical special economic zone (SEZ) and the process involved, the following layout has been planned for the proposed plant (Figure 8.2).

The plant is proposed to be built on a typical SEZ area of 5000 m². It enjoys a compact design including both raw material and inventory shipping as well

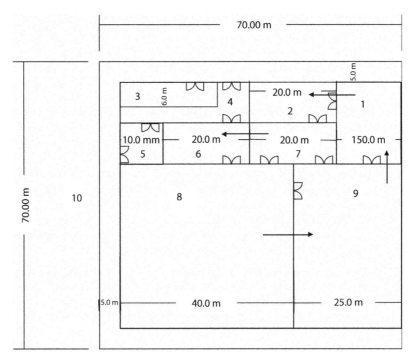

Figure 8.2 Layout of the plant: (1) Shipping area, (2) raw material input, (3) canteen, (4) office, (5) restroom, (6) gowning room, (7) quality control, (8) dialysis fiber spinning area, (9) cartridge potting area, and (10) corridor. (All dimensions are in meters.)

as production facilities. The plant has four distinct areas of activity, and they are (a) raw material and input goods sections (marked 1 and 2); (b) manufacturing and quality control (marked 7 and 8); (c) the packaging and shipping section; and (d) the administrative and official workplace as well as canteen and restroom for employees (marked 3, 4, and 5).

The logical process in the development of the layout is as given:

1. The process of manufacturing dialysis fibers has been upscaled with the following equation[1]:

$$N = \frac{T * P}{60 * D * E} \tag{8.1}$$

where

P is the required output rate (units of output/period)
T is the processing time per unit (minutes)
D is the duration of operation (in hours)
E (0.6) is the efficiency of the machine
N is the number of machines operated

191

The value of each variable is important. The process discussed in Chapter 3 is capable of spinning 1200 fiber units (each fiber unit is of a standard dialyzer length, i.e., 30 cm) of dialysis hollow fibers in 4 hours' time (by a single spinning device). Hence, T is $(4 * 60)/1200 = 0.2$ minutes. D is the available time in a day, which is taken to be 16 hours. There is no official documentation on the number of cartridges that a dialysis plant produces, but data suggest that around 30,000 cartridges per month are a minimum for any standard plant manufacturing dialyzers. Thus, with these values, it is clear that around 1,000 cartridges will be manufactured per day and if each contains 10,000 fibers, then around 10,000,000 fibers will be spun per day. Hence, N, or the number of machines, comes out to be around 3400. This is very important data as this decides the availability of space, and it has been determined that 1600 m^2 is needed for setting up 3400 such units. The whole area of fiber spinning and potting is a clean room area (class 10,000).

2. A logical and continuous material flow pattern is maintained in the layout, such that minimum interruptions are caused. The arrows in the layout indicate the material flow around the plant. The quality control and assurance sections are located such that both incoming goods to manufacturing and outgoing finished products go through strict quality checking.

3. Provisions for canteens and restrooms have been made for employees. A wide corridor has been provided for evacuations during fire hazards or any unforeseen circumstances.

4. The production rate has been planned so that a slow continuous ramp-up is possible, which can go on in tandem with the sales and other activities of a growing company. As discussed in the previous section, there are four types of products that the present technology can manufacture. Hence, it has been proposed that production will commence within 7 months of commissioning of the plant. This is because the first 6 months would go into synchronizing the various operations involved in production, recruitment of staff, procurement of miscellaneous items and machineries, and so on. The number of cartridges produced during the period of 7–10 months would be 850 for each of the four types. This is envisaged to grow steadily, reaching 5,100 cartridges of each type by the end of the second year and then a steady 8,500 cartridges from the 29th month, which totals the target production capacity of 30,000 cartridges per month.

8.4 Business Plan, Key Activities, and Key Resources

Figure 8.3 illustrates the business plan envisaged for commercializing the product. Renalyx Health Systems Pvt. Ltd. is the industrial partner for the

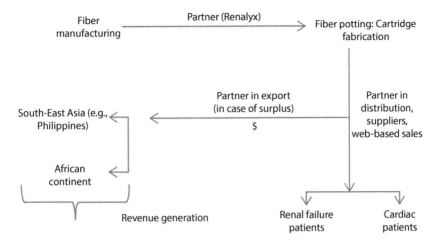

Figure 8.3   Business model developed for the product.

invention. Fiber spinning and cartridge preparation have been planned and sales are to be undertaken by partnering with existing distributors. Direct web-based sales to customers are also an option that will be explored. The preliminary investments are for these activities. Government hospital supplies will be at subsidized rate contracts. The detailed financial analysis for the products and the business model proposed are discussed in the ensuing section.

8.5 Financial Analysis

The market demand for cartridges surpasses the anticipated production capability of the plant mentioned earlier by a great extent. The present business and financial models are capable of producing 30,000 cartridges per month or 3.6 lakh cartridges per annum. This is roughly 4% of the anticipated demand. However, if the financial model is viable for one production unit, then this model can be replicated. Table 8.1 depicts the start-up fund requirements for the plant.

The plant establishment cost is Rs. 7,000,000 as per SEZ norms. The spinning device designed is estimated to cost around Rs. 5,100 each, and, since 3,400 cartridges are required, an investment of around Rs. 17,500,000 is necessary. The potting device and splicing device can be custom-made and are estimated to cost around Rs. 500,000 each. The quality assurance/quality control (QA/QC) equipment consists of a scanning electron microscope to measure the dimension of fibers and surface morphology visualization. A tensile strength–measuring device to maintain consistent fiber mechanical strength, a porometer to determine the porosity of fibers, and a rheometer to determine the consistency of the spinning solution are also required. The finances for office furniture and production furniture requirements are in the

193

Table 8.1 Start-up Fund Requirements (INR)	
Plant establishment	7,000,000
Spinning devices (3400)	17,500,000
Potting device	500,000
Slicing device	500,000
Packaging machine	2,000,000
QA/QC equipment	20,000,000
Office furniture	2,000,000
Production furniture	5,000,000

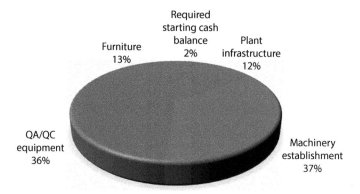

Figure 8.4   Fund requirements for the plant.

order of 20 and 50 lakhs, respectively. The respective percentages of start-up fund requirements are presented in Figure 8.4. It is interesting to note that the machinery and QA/QC equipment occupy 37% and 36% of the fund requirements, whereas the starting cash balance is only 2%, which is enough to start the manufacturing process.

Figure 8.5 depicts the funds required for the plant set-up. The three sources of funds expected are government grants, soft loans from banks, and investment from venture capitalists. The government fund granting agencies are the Biotechnology Industry Research Assistance Council (BIRAC), the Department of Science and Technology (DST), the Department of Biotechnology (DBT), and the Scientific and Engineering Research Board (SERB). The second type of funding is soft loans from the Small Industries Development Bank of India (SIDBI). Venture capitalists are the final section of investors who will be approached for fund-raising. It is to be noted that two soft loans of equivalent sums (Rs. 1 crore) are planned to be raised from

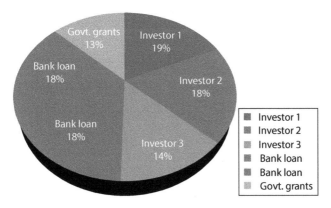

Figure 8.5   Source of funds for the plant.

two banks, whereas two sums of Rs. 1 crore are planned to be raised from two investors, and Rs. 75 lakhs from a third one. A grant of Rs. 75 lakhs is to be sought from the government. The respective percentages are presented in Figure 8.5.

Once the plant gets commissioned, the early months are usually the "numb" stages when there is no production. The assumption that has been taken into account in this model is that the initial commissioning of the production line, including spinning, as well as those for potting and splicing, will take about 2–3 months. Moreover, quality control equipment would require training, operation, and installation time. To have a conservative estimate, it is assumed that the first 6 months will experience no production. Slowly, a ramp-up is expected, as discussed previously. The success of any company is almost solely dependent on the sales of the product. This also determines the capability of a company to drive the market. Keeping in mind the existing competition and understanding that availability of dialyzers is also an issue in India, it is planned that the sales will be concentrated in specific areas for the start-up. It is expected that from the seventh month 3440 cartridges will be manufactured (850 cartridges of each type (HE, HPM, HCO, and BO)). This production will be ramped up from the 10th month to 6800 cartridges per month, which will increase to 10,000 by the end of the first year and 20,000 by the end of the second year, ultimately reaching the target of 34,000 cartridges by the end of 3 years. Since the process of manufacturing for each of the products remains the same, the number for each type remains the same. Sales, however, is not just dependent on production. It is primarily dependent on entering a very competitive and already existing market and on carving a niche for the developed product. This requires a very aggressive marketing campaign, both print and online. All these investments also have been taken into account, as operating expenditures, and, based on these, the annual sales forecast shows (Figure 8.6) that a steady ramp-up is expected per annum. However, if the brand becomes established then sales would increase much rapidly than what is depicted.

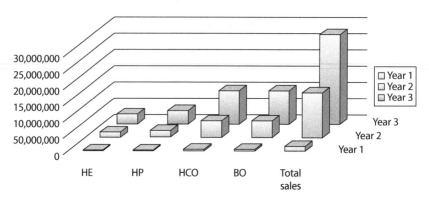

Figure 8.6   Annual sales forecast.

Based on sales and several other factors, the gross margin and net profit are shown in Figure 8.7. At this juncture, it is imperative to introduce "several other factors," which would in turn throw light on the reality and depth of analysis. One of the elements hidden in these factors is the operating expense. These expenses include high budget heads such as accounting, advertising, insurance, legal professional payment, license renewal to medium budget heads such as salaries, utilities, travel, telephone bills, meals and entertainment, and website maintenance. High budget heads would experience annual expenses as high as 10 crores (advertising) and medium budget heads would experience annual expenses of around 6 lakhs (telephone). This takes care of the operating expenses. However, the profit is basically dependent on the cash flow of the company. Cash flow is dependent on the difference between cash receipts and cash disbursements. While earning from sales comes under receipts, operating expenditure comes under disbursements, along with other factors such as loan interest repayment, loan principal repayment, and income tax. The balance between the two represents the cash flow and the difference between the two represents the net cash flow. Once the net cash flow becomes positive, the breakeven point is reached.

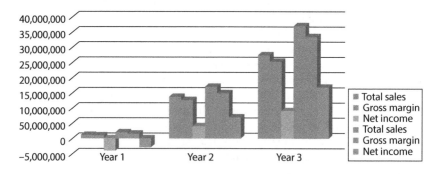

Figure 8.7   Total sale, gross margin, and net profit.

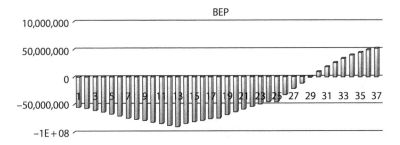

Figure 8.8   Breakeven point analysis.

Keeping the preceding discussion in mind, it can be observed from Figure 8.7 that the net income is negative during the first year of production. However, as sales increase, the cash flow increases, which slowly increases the gross margin and net income. From Figure 8.8, it is evident that the breakeven point is reached at around 30 months. After that, the business becomes self-sustaining. The gross margin and net income increase, and the company is expected to be cash-rich from the third year of operation. With this business and financial model, it is calculated that the cost of each dialyzer manufactured is Rs. 168 and the selling price is envisaged to be Rs. 300 (including distribution channels and profit). With proper distribution, not only would these cartridges be available in the Indian market at the said price, but a consistent and immediate end user base would also be available.

8.6 Prototypes Developed

The discussions in the preceding sections reveal that the indigenous cartridge cost would be Rs. 300. To test the feasibility of manufacturing cartridges from spun membranes, a suitable manufacturer was identified with the capability to pot them. The technique involved is very difficult, and presently there is only one company in India possessing the facility to do it. With the help of the facility, the prototype of a cartridge was made (as illustrated in Figure 8.9; placed alongside the commercially available one). The geometric, physical, and performance characteristics of developed fibers are presented in Table 8.2 along with those of Fresenius fibers. It can be observed from this table that indigenously prepared fibers closely match the commercial counterparts on various clinical, biological, and physical parameters, but at a fivefold less cost.

8.7 Conclusion

It is a crucial aspect of engineering to analyze the feasibility of a technology in actually reaching the end user. The scalability of a particular technology is primarily dependent on the economics of production. It is important for an engineer to carry out a complete business and financial model analysis

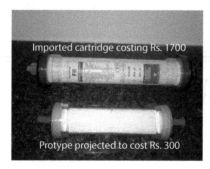
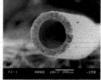
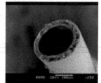

Fresenius F 60   Fibers from the present work

Figure 8.9 Prototype developed in this work alongside commercially available cartridge.

Table 8.2 Comparison of Characteristics of Developed and Commercial Fibers

Characteristics	Commercial Fresenius F60 Hollow Fibers	Fibers Spun from This Work
Inner diameter (μm)	200	220
Thickness (μm)	35–40	35–40
Biocompatibility	√√	√√
Toxin removal capabilities	√√	√√
Breaking stress (MPa)	11.5	12.7
Target patients	Renal	Renal/cardiac/myeloma
Price (Rs.)	1500–2000	300

in order to find out the cost of the product and the price of the same. It was seen in this chapter that the complete technology, from polymer to fiber to cartridge, requires a sound business and financial model to be scaled up along with an industrial partner. Importantly, what this chapter reveals is that once a manufacturing unit has been set up and production has commenced, it takes only Rs. 168 to manufacture a dialysis cartridge. Hence, the imported version of the same is available to the general public at 10 times the cost. However, with a proper business model and with sound distribution in place, with this technology, a cost as low as Rs. 300 can be realized, that too by keeping everyone satisfied, starting from the end users to the distributors. This is the true sign of a feasible product.

Reference

1. Arora, K.C. 2004. *Production and Operations Management.* Firewall Media, New Delhi, India.

Conclusion: Future Trends and Scope

Where the mind is without fear and the head is held high.

—**Rabindranath Tagore**

9.1 Future Trends in Dialysis Research

Future trends in dialysis research are not likely to be bound within the realms of classical membranes. Better and more biocompatible membranes, which form a staple source of research for engineers and scientists, are being developed; moreover, innovative options for "membrane-less" dialysis as well as options that could mimic human nephrons more closely are also being researched.

1. *Next-generation biocompatible membranes*: Next-generation biocompatible membranes that have an enhanced ability to reduce oxidative stress are being developed. Anti-blood coagulation too is an important aspect, since it reduces the use of heparin during dialysis, thereby bringing down treatment costs. Achieving higher filtration rates by engineering larger, yet appropriate pore sizes with sharp cutoffs is the target for next-generation membranes. Apart from this, other aspects such as fiber bundling, minimizing pure water usage as dialysate, and developing portable and wearable dialyzer units (as belts) are also possible avenues of research.

2. *Innovative membranes*: Edward Leonard's group[1] proposed microfluidic principles to carry out dialysis between two streams, and the Nissensson group[1] developed a novel replacement device called

the "human nephron filter." The former group discusses two liquid streams coming in contact with each other at a low Reynolds number, thereby inhibiting mixing and thus maintaining two distinct phases. Mass transfer occurs from one stream to the other and they can be separated downstream. The latter group discusses a device mimicking a glomerulus and a tubule and has two filters, the G membrane (mimicking glomerulus) and the T membrane (mimicking tubule). There is no use of dialysate and, theoretically, 1/10th of 1 m^2 surface area is required. Nanotechnology has been utilized to achieve the pores of the two membranes, with the membranes being one molecule thick. Silicone nonporous membranes are being developed with the help of the sacrificial layer technique.[1] Nature-inspired slit-shaped pores are also being explored as an alternative to round pores as they have the potential for better performance.

3. *Living membranes*: It is well known that dialysis undertaken three times per week barely provides 15–20 mL/min of equivalent clearance. Scientists, nephrologists, and engineers are working together to approach ideal kidney functioning with possible wearable devices that would function continuously to reject uremic toxins. Ongoing research, specifically to address this problem, has paved the path for endothelial cell-seeded microfabricated capillary network devices.[1] Furthermore, cells cultured from adult human kidneys are grown in the inner surface of hemofiltration membranes. This helps in maintaining blood homeostasis and in reabsorbing selective solutes during filtration.

9.2 Water Requirements and Standards

Water plays an important role in dialysis simply due to the sheer volume of its use. The dialysis treatment for a single patient requires about 18,000 L/week.[2] The Association for the Advancement for Instrumentation (AAMI) and European Pharmacopeia (EP) are the two most general standards that are followed worldwide.[3–5] The basic difference between these two standards is that although both have similar stringent limits for chemical contamination, the level of stringency differs when it comes to microbial contamination. The levels are summarized in Table 9.1.[6] It is evident that EP standards are more stringent than AAMI standards for microbial contamination. The question, however, is whether it is worthwhile to achieve such levels of purity for dialysate fluids.[7]

Other than the standards discussed, the International Organization for Standardization (ISO) has its own water-quality standards for dialysis. Although ISO 26722:2009 states the standards pertaining to water-quality equipment,[2] ISO 13959:2009 and ISO 13958:2009 deal with the quality of water produced for dialysis. The microbiological levels of contamination are

Table 9.1	Levels of Allowable Microbial Contamination	
Standard	AAMI	EP
Microbial contamination	Upper level—200 c.f.u./mL	Less than 100 c.f.u./mL
	Disinfection, retesting to be performed above 50 c.f.u./mL	
Endotoxin contamination	Upper level—2 IU/mL	Less than 0.25 IU/mL
	Disinfection, retesting to be performed above 1 IU/mL	

also mentioned in ISO 13958:2009.[2] However, some ambiguity still exists regarding the allowable concentration levels of certain inorganic chemicals such as nitrate, and, importantly, none of the current standards addresses the allowable levels of organic chemicals.[2] However, it is pretty much evident that water quality as well as quantity play a vital role in dialysis and thorough knowledge about reverse osmosis operation and membranes goes a long way in addressing issues at hand on the spot. Keeping such requirements under consideration, designs for the development of resins that can adsorb the washed toxins from the dialysate side are being sought,[8] which will thereby decrease the requirement for fresh water. This could potentially bring down treatment costs and would help in doing away with costly reverse osmosis units.

9.3 Dialyzer Reuse and India

Once the membranes have been synthesized and spun into clinical grade fibers, they are packed to form a dialyzer cartridge as discussed in previous chapters. The formation of a dialyzer cartridge is a science in itself and requires volumes of material to acquire mastery over the art. This is out of the scope of the present book, but readers may refer to a host of materials available on this subject.[9–13] For understanding the commercial aspect of a technology, it is important to research on its possible reuse. Reuse plays an important part in the economics of production, simply because if a material is bought for $1 and used 10 times, its effective cost is $0.1 per use (neglecting associated costs). For a developing economy, like South-East Asia, this is an important consideration. The general mind-set of the consumer would be to reuse the technology as many times as possible, so as to decrease the "per use" costs involved. This seems like "an affordable solution," which in reality it is not! This is because, for each reuse, the performance of the technology gets depleted by a definite percentage, as per the laws of nature (or thermodynamics). There is no ideal process that ensures the complete recovery of performance after each usage, and hence there is a substantial decrease in efficiency. The history of dialyzer reuse is almost as old as the technology itself and makes for an interesting read.

9.4 Dialyzer Reuse: Steps

The basic steps for cleaning hollow-fiber dialyzers are (1) cleaning, (2) testing, (3) sterilization, and (4) storage.

1. *Cleaning*: This involves two methods. One is the flushing of water through the blood compartment of the hemodialyzer and the other is reverse ultrafiltration (RUF). RUF is particularly efficient in cleaning high-flux dialyzers.[14] It involves forcing water from the dialysate into the blood compartment by applying pressure, which finally exits the blood ports.[15] The important thing is to prevent fiber collapse during this process due to excessive pressure gradients. Cleaning solutions typically comprise dilute bleach solutions, sodium hydroxide, and hydrogen peroxide.[16] Bleach solutions are inexpensive and are pretty good at removing fouling from membranes. However, when used on polysulfone membranes, blended with polyvinylpyrrolidone (PVP), there is an unusual increase in ultrafiltration rates during subsequent dialysis therapies.[9] This is due to leaching of PVP from the membrane matrix, leading to pore restructuring, and increase in pore size, leading to high loss of albumin.[17,18] Other than bleach, no other cleaning agent can sustain itself due to various problems. In this regard, peracetic acid (PAA) proved to be a blessing in disguise, as it can clean as well as sterilize.[19] PAA is sometimes mixed with hydrogen peroxide and acetic acid.[18] The resulting solution is sold in a concentrated form and diluted with water before usage, which breaks down the solution into acetic acid, oxygen, and water. Sterilization with PAA also requires testing the potency of the chemical prior to usage for dialysis therapy.

2. *Testing*: Two objectives are sought in this mode. One is the efficiency of the dialyzer post cleaning and the other is testing the membrane integrity. It is generally agreed within the nephrology community that a 10% decrease in urea removal capacity of dialyzers is generally acceptable for reprocessed or reused hemodialyzers. Dr. Gotch[20] devised a simple technique to estimate the total cell volume (TCV) of a hemodialyzer post treatment. This was done by emptying the contents of the blood side of a dialyzer into a graduated cylinder by pressurizing one of the blood side ports using inflation bulbs. For high-flux dialyzers, the dialysate ports are closed since high membrane permeability can lead to errors generated during pressurization. This forms the basis of TCV measurements, and for subsequent reuse the same methodology can be followed. A 20% decrease in TCV corresponds to a 10% decrease in urea removal capacity.[20] Membrane integrity is tested by applying pressure and measuring the pressure decay rate across membranes.[18]

3. *Sterilization of membranes*: Sterilization of membranes is preceded by removal of air from the dialyzer, which is achieved by continuous

and adequate flushing with water, making the dialyzer air-free and filled with fluid. The most common microorganisms causing concern in present-day hemodialysis units are Gram-negative bacteria and non-tuberculosis mycobacteria[21,22] as well as endotoxemia.[23] The AAMI recommended that the water used to prepare disinfectant solutions should contain less than 1 ng/mL of endotoxin. The very first disinfectant that was commercialized was formalin (37% formaldehyde with methanol). This suffered a major setback due to its environmental hazards. As discussed previously, PAA is an important sterilizing agent that was first used as a germicide. It is a very strong oxidizing agent and is diluted to minimum 4% by volume. It was also discussed previously that all such sterilizers have to be tested for their presence within the dialyzer before being reused. Schiff reagents are used to test for the presence of formalin, whereas strong oxidizing agents like PAA require test strips.[18] Another method to sterilize dialyzers that has gained popularity over the years is thermo-critic reprocessing.[18] This involves heating dialyzers at 100°C–120°C in an oven for 20 hours. This does not require any chemical treatment and in some ways promoted the popularity of the polysulfone family due to its excellent thermal stability. However, thermo-critic reprocessing also posed a different challenge and that is differential expansion of the housing and potting material. Unequal expansion of the two could lead to gaps through which blood could leak into the dialyzer compartment.[18] This was solved by adding 1.5% citric acid to water, which brought down the process temperature to 95°C.

4. *Storage*: Another aspect that needs to be understood is the "dwell time." This involves proper storage time for specific sterilization techniques chosen, for example, PAA-sterilized dialyzers require 11 hours at room temperature.[18]

There are automatic reprocessing machines that not only automatically clean dialyzers, but also record the patient's name and the dialyzer model, automatically label, and accommodate multiple ultrafiltration rates and volumes of dialyzers.

It is clear that reprocessing of hemodialyzers is an important operation. But, in the Indian or, for that matter, any developing country's context, it is very difficult to afford such reprocessing machines at each and every dialysis center. A typical hemodialysis machine itself costs as high as $9000, a reverse osmosis water treatment plant costs $1500, and other paraphernalia too cost $1500. Importantly, it is well reported that for dialysis therapies, around 58% of the total cost per dialysis session can be attributed to the cost of the dialyzer alone.[24] This is because, although the infrastructure for dialysis sessions is provided and subsidized by the government, it is the patient's responsibility to bring a dialyzer for treatment. In India, all dialyzers are imported and it becomes cumbersome and at times frustrating for patients to get hold of a new dialyzer; hence, crude methods are adopted for cleaning used ones and reusing them again. As such, for developing countries, the problem lies not in the paraphernalia, but in the economics of manufacturing a dialyzer.

9.5 Holistic Overview

In this book, hemodialysis and the current scenario in developing countries such as India have been explored thoroughly, and a possible solution to address the predicament of renal failure patients has been attempted. The dual problem of unaffordability and nonavailability of hemodialyzers in the Indian market places the patients in a precarious position, more often than not resulting in premature death. Elsewhere in the world, even on dialysis, people survive for decades with a healthy lifestyle. The inability to manufacture the basic fundamental unit of a dialyzer, that is, a hemodialysis grade hollow-fiber membrane, is a major bottleneck for most dialyzer companies around the world as well as for the indigenous development of dialyzers.

In this work, a complete technology has been invented, perfected, and optimized for spinning hemodialysis fibers. This includes polymer selection, biocompatibility evaluation, indigenous low-cost spinning technology development, as well as postprocessing technologies to synthesize market-competitive fibers.

During the entire development process, a pragmatic outlook has been adopted, by maintaining close interaction with hospital and industrial partners. This has helped in achieving better results in terms of fiber spinning, and critical issues like polymer selection and maintenance of clinical dimensions have also been addressed.

It is important for an engineer to test the feasibility of the innovation by understanding the economics of production and to come out with a feasible plan to take the product to the end user. This has also been kept in mind, and business and financial plans have been formulated to help take the product to the end user.

The present work is a conglomeration of research, innovation, and engineering. Individually, they are three different genres, but merging them has resulted in a prototype. Once the prototype undergoes rigorous financial and business analysis resulting in its manufacturing and a consumer market, it becomes a product. The pathway of an innovation from the lab to the market is perhaps the most challenging and tortuous route ever encountered. What is more challenging is that there is no fixed or even predictable formula for success. Starting from theory to proof of concept to prototype to ultimate manufacturing may take decades of effort (Figure 9.1). However, it is important to remember and recount the adage "failure is the pillar of success."

India has the potential to become a manufacturing giant with various innovations finding their way from the lab to the market. In this process, it is important to engage in socially relevant research generating wealth, providing employment, and ensuring sustainability. This is only possible with an entrepreneurial mind-set and with a fearless attitude to try out new things, by delving into the unknown.

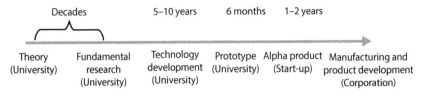

Figure 9.1   From lab to commercialization.

References

1. Humes, H.A., Fissell, W.H., and Tiranathanagul, K. 2006. The future of hemodialysis membranes. *Kidney Int.* 69: 1115–1119.
2. Hoenich, N.A., Levin, R., and Ronco, C. 2010. Water for haemodialysis and related therapies: Recent standards and emerging issues. *Blood Purif.* 29: 81.
3. 85AAMI Standard and Recommendation Practices. 2000. AAMI/DS-1 RD62. Water treatment equipment for hemodialysis applications. Association for the Advancement of Medical Instrumentation, Arlington, VA, pp. 2–32.
4. Baz, M., Durand, C., Ragon, A., Jaber, K., Andrieu, D., Merzouk, T., Purgus, R., Olmer, M., Reynier, J.P., and Berland, Y. 1991. Using ultrapure water in hemodialysis delays carpal syndrome. *Int. J. Artif. Organs* 14: 681–685.
5. European Phamacopeia, 3rd edn, Supplement 2001: Monograph 1997:1167 (corrected 2000). Haemodialysis solutions, concentrated, water for diluting.
6. Pontoriero, G., Pozzoni, P., Andrulli, S., and Locatelli, F. 2003. The quality of dialysis water. *Nephrol. Dial. Transplant.* 18(Suppl. 7): 21–25.
7. Tielemans, C., Hoenich, N.A., Levin, N.W., Lonnemann, G., Favero, M.S., and Schiffl, H. 2001. Are standards for dialysate purity in hemodialysis insufficiently strict? *Semin. Dial.* 14: 328–336.
8. Winchester, J.F. and Ronco, C. 2010. Sorbent augmented hemodialysis systems: Are we there yet? *J. Am. Soc. Nephrol.* 21(2): 209–211.
9. Vander Velde, C. and Leonard, E.F. 1985. Theoretical assessment of the effect of flow maldistributions on the mass transfer efficiency of artificial organs. *Med. Biol. Eng. Comput.* 23: 224–229.
10. Ronco, C., Brendolan, A., Crepaldi, C., Rodighiero, M., and Scabardi, M. 2002. Blood and dialysate flow distributions in hollow-fiber hemodialyzers analyzed by computerized helical scanning technique. *J. Am. Soc. Nephrol.* 13: S53–S61.
11. Ronco, C., Scabardi, M., Goldoni, M., Brendolan, A., Crepaldi, C., and La Greca, G. 1997. Impact of spacing filaments external to hollow fibers on dialysate flow distribution and dialyzer performance. *Int. J. Artif. Organs* 20: 261–266.
12. Ronco, C., Brendolan, A., Crepaldi, C., Rodighiero, M., Everard, P., Ballestri, M., Cappelli, G., Spittle, M., and La Greca, G. 2000. Dialysate flow distribution in hollow fiber hemodialyzers with different dialysate pathway configurations. *Int. J. Artif. Organs* 23: 601–609.
13. Poh, C.K., Hardy, P.A., Liao, Z., Huang, Z., Clark, W.R., and Gao, D. 2003. Effect of spacer yarns on the dialysate flow distribution of hemodialyzers: A magnetic resonance imaging study. *ASAIO J.* 49: 440–448.
14. Ward, R.A. and Ouseph, R. 2003. Impact of bleach cleaning on the performance of dialyzers with polysulfone membranes processed for reuse using peracetic acid. *Artif. Organs* 27(11): 1029–1034.

15. Varughese, P.M. and Andrysiak, P. 2000. Dialyzer reuse. In: *Dialysis Technology: A Manual for Dialysis Technicians*, 2nd edn., Curtis, J. and Varrughese, P. (eds.). National Association of Nephrology Technicians/Technologists, Dayton, OH.
16. Dennis, M.B., Vizzo, J.E., Cole, J.J., Westendorf, D.L., and Ahmad, S. 1986. Comparison of four methods of cleaning hollow-fiber dialyzers for reuse. *Artif. Organs* 10(6): 448–451.
17. Graeber, C.W., Halley, S.E., Lapkin, R.A., Graeber, C.A., and Kaplan, A.A. 1993. Protein losses with reused dialyzers. *J. Am. Soc. Nephrol.* 4(3): 349.
18. Azar, A.T. (ed.). 2013. *Modeling and Control of Dialysis Systems*, Vol. 1: Modeling Techniques of Hemodialysis Systems, SCI 404. Springer, Berlin, Germany.
19. Berkseth, R., Luehmann, D., McMichael, C., Keshaviah, P., and Kjellstrand, C. 1984. Peracetic acid for reuse of hemodialyzers, clinical studies. *Trans. Am. Soc. Artif. Intern. Organs* 30: 270–274.
20. Gotch, F.A. 1980. Mass transport in reused dialyzers. *Proc. Clin. Dial. Transplant Forum* 10: 81–85.
21. Band, J.D. and Fraser, D.W. 1979. Peritonitis caused by a Mycobacterium chelonei-like organism (MCLO) associated with chronic peritoneal dialysis (CPD) [abstract no. 214]. *Program and Abstracts of the 19th Interscience Conference on Antimicrobial Agents and Chemotherapy*. American Society for Microbiology, Washington, DC.
22. Petersen, N., Carson, L.A., and Favero, M.S. 1981. Bacterial endotoxin in new and reused hemodialyzers: A potential cause of endotoxemia. *Trans. Am. Soc. Artif. Intern. Organs* 27: 155–160.
23. Peterson, N.J. 1983. Microbiologic hazards asociated with reuse of hemodialyzers. In: *Proceedings of the National Workshop on the Reuse of Consumables in Hemodialysis*, Sadler, J. (ed.). Kidney Disease Coalition, Washington, DC, pp. 119–134.
24. Lobo, V., Gang, S., Shah, L.J., Ganju, A., Pandya, P.K., Rajapurkar, M.M., and Acharya, V.N. 2002. Effect of reuse of hollow fiber dialyzers upon Kt/V(Urea): A prospective study. *Indian J. Nephrol.* 12: 40–46.

Index

Index